There's No Place Like Home Video

VISIBLE EVIDENCE

Edited by Michael Renov, Faye Ginsburg, and Jane Gaines

VISIBLE EVIDENCE, VOLUME 12

There's No Place Like Home Video

James M. Moran

University of Minnesota Press

Minneapolis

London

Published by the University of Minnesota Press
111 Third Avenue South, Suite 290
Minneapolis, MN 55401-2520
http://www.upress.umn.edu

Library of Congress Cataloging-in-Publication Data

Moran, James M., 1962–
    There's no place like home video / James M. Moran.
        p.   cm. — (Visible evidence ; v.12)
    Includes bibliographical references and index.
    ISBN 0-8166-3800-4 (alk. paper) — ISBN 0-8166-3801-2 (pbk. : alk. paper)
    1. Video recordings.   2. Amateur films—History and criticism.   3. Video
  recordings—Production and direction.   I. Title.   II. Series.
    PN1992.945 .M67 2002
    384.55'8—dc21

                                                                    2002000553

Printed in the United States of America on acid-free paper

The University of Minnesota is an equal-opportunity educator and employer.

12 11 10 09 08 07 06 05 04 03 02   10 9 8 7 6 5 4 3 2 1

This book is dedicated to my father, John Moran,
to my elder siblings, Jack, Ellen, and Cathy,
to my grandmother Florence,
and to the memory of Marilyn, my mother.

Contents

Acknowledgments

Without the support of the following people, this book would still be the formless musings of a well-meaning but callow graduate student. I am most indebted to my adviser and mentor, Michael Renov, for his academic guidance and unflagging encouragement to produce my best work. I am also deeply grateful to Marsha Kinder, David James, Lynn Spigel, Tara McPherson, and Lee Stork for the intellectual stimulation and opportunities for professional development they provided me at the University of Southern California School of Cinema–TV. And for making this publication possible, I must thank Jane Gaines at Duke, Vivian Sobchack at UCLA, Nancy Vickers at Bryn Mawr, and Jennifer Moore at the University of Minnesota Press.

Among my fellow classmates at USC, who have since pursued careers all over the map, I want to make special mention of Steve Anderson, Clark Arnwine, Harry Benshoff, Rich Cante, Pamela Chandran, Donna Cunningham, Eric Freedman, Janice Gore, Sean Griffin, Vicky Johnson, Mary Kearney, Tom Kemper, Angelo Restivo, Luisa Rivi, Bhaskar Sarkar, Valentin Stoilov, Alison Trope, Cristina Venegas, Karen Vered, Karen Voss, Dan Walkup, and Holly Willis. It has been my honor and privilege to spend countless hours laughing and debating with these colleagues, the best and the brightest of my acquaintance.

Finally, to my friends whose emotional support pulled me through the trials of doctoral candidacy, I wish the best of success and happiness in life, as they have given me: Nuncle, Randy, Bishnu, Catherine and Cliff, Afzal, Phil, Alex, Elizabeth, Mushtaque, Shantanu, Nahid and Linda, Rick and Joe, Jim and Derek, Sandie, Patti, Frank, Susan, Chris and Linda K., Hiro and Velvet, Paul and Karina, Flash and Tina, and other old buddies and new acquaintances too numerous to mention here.

To Nuncle)
with
Love)

Jim

Introduction
Medium Theory, Home Video,
and Other Specifications

> *On a theoretical level, I think it is no longer useful or even*
> *pertinent to treat photography as a thing in itself, or cinema*
> *as an ontology, or video as a specific medium. . . . I think that*
> *we have never been in a better position to approach a given*
> *visual medium by imagining it in light of another, through*
> *another, in another, by another, or like another. Such an*
> *oblique, off-center vision can frequently offer a better*
> *opening onto what lies at the heart of a system.*
> :: Philippe Dubois, "Photography *Mise-en-Film:* Autobiographical
> (Hi)stories and Psychic Apparatuses"

When, in 1957, André Bazin first posed the question "What is Cinema?"
his self-conscious inquiry into the specificity of the medium epitomized the
fundamental discourse of several generations of classical film theory from
Hugo Münsterberg to Christian Metz. Of course, as cinema evolved from
its humble beginnings as a peep show curiosity into a commercial institution with mass appeal, individual theorists posed Bazin's question from
perspectives inflected by their own aesthetic preferences and the historical
moment at which the invention of certain technologies constituted cinema's
"basic apparatus." Inevitably, therefore, debates over cinema's "proper"
constitution raged from decade to decade, often splintering film theory into
opposing camps. Nevertheless, despite variations of agenda, such as Béla
Balázs's entrenchment of cinema within the paradigms of Russian formalism,
or Siegfried Kracauer's insistence on film's inherent predilection for photographic realism, film theorists from the turn of the century into the 1970s
shared an elemental pursuit: to identify and define the essence of cinema
as an autonomous medium of artistic production. Implicitly adopting and
adapting the specificity thesis of Gotthold Ephraim Lessing's *Laocoön*,
which first advanced that each art form, owing to the specific structure of

its physical medium, should be differentiated from all other art forms according to the formal effects to which it is best suited, classical film theorists centered their interrogations on cinema's distinctions from its aesthetic ancestors. What exactly, they asked, is this medium that appears at once novel and yet a synthesis of photography, theater, music, dance, and literature? What is cinema's unique identity? What should constitute a purely cinematic practice?

Although the predominance of the specificity thesis within classical film theory was overdetermined by various causes, such as the increasing influence of modernism in the arts and the desire by enthusiasts to raise the "movies" from their lowbrow status, it is perhaps the cinema's highly complex technological apparatuses of production and exhibition, unlike those of any medium before it, that inspired so many detailed examinations of its inherent properties, potentials, and limitations. Cinematic production has been subjected by classical film theorists to a set of "techno-aesthetic" distinctions: that is, they tended to define cinema first by identifying the primary technological constituents of its "basic apparatus" (i.e., lens, film stock, screen, projector) and second by theorizing the manner in which this technology determines the possible aesthetic effects of individual films themselves. Although this approach would reap precise and detailed accounts of cinematic practices ranging from Sergei Eisenstein's taxonomies of montage to Metz's Grand Syntagmatique, it has inevitably led to theoretical roadblocks.

Perhaps the "Materialtheorie" of Rudolf Arnheim best represents the drawbacks of a specificity thesis grounded in a techno-aesthetic paradigm. In *Film as Art,* Arnheim developed a theoretical model of film derived from the predominant cinematic form at the time he was writing: silent black-and-white images framed within a 1.33:1 aspect ratio. Fixing this temporary technological configuration as an ontology of cinema's material limitations, whose narrow exploitation constituted a "true" cinema practice, Arnheim postulated an a priori system of aesthetic values against which he measured the "purity" of individual films, by definition eliminating color, sound, and wide-screen technologies, whose arrival after the fact of his model ensured their exclusion. By failing to acknowledge that the foundation of his aesthetic system had been determined by the technologies available at one historical moment, Arnheim retreated to a restrictive brand of prescription, denying these new technologies any prospect of enhancing or creating great cinematic art and nostalgically pining for a definition of cinema constricted to "a unique experiment in the visual arts which took place in the first three decades" of the twentieth century.[1] A tacit technological determinist, Arnheim declared that visual recording mechanisms had reached their apotheosis during the early twenties, as if they had been invented expressly for the production of "silent films" per se, and that the introduction of new devices must inevitably undermine cinema aesthetics. Unlike Bazin, who in "The Myth of Total Cinema" considered the possibility that the cinema had not yet been completely invented, because film technology was limited only by accidents of science and economics rather than by the imagination of its inventors,[2] Arnheim mistook cinema's historical existence with eternal essence and prescribed a film aesthetic determined more by technology than by human intention. Indeed, although Bazin himself frequently advocated an ontology of film realism, his question "What is Cinema?" may be asked rhetorically, even asymptotically, since an answer must always be deferred over the ongoing course of its development. That is, the technological specificity of the filmic medium can never be stabilized as a single synchronic configuration outside of the diachronic fluctuations of its history.

It is precisely this notion of medium specificity—one that rejects the ahistorical determinism, inherent properties, and prescriptive either/or binarisms of Arnheim—upon which the central tenets of this study will be founded. Here, however, unlike in Bazin's work, the fundamental question of my inquiry will be "What is Video?" and my investigation will forgo pinning down an answer within a techno-aesthetic paradigm. One might argue that the quest for video's technological and aesthetic specificity over the last three decades according to the paradigms established by classical film theory has led not only to outlandish conclusions, such as Fredric Jameson's notorious declaration that there will never be a theory of video, but also to a decline in the pursuit of medium specificity itself.[3] Certainly, the rise of alternative schools of thought since the 1960s has altered in part the course of media analysis away from apparatus theory: feminism, Marxist cultural studies, and postcolonial critique, for example, while based on various epistemological tenets, share methodologies focused more on the ethnographic and sociopolitical aspects of media practice than they do on medium specificity. Nevertheless, like Jacques Derrida's "gram," which defers the prospect of self-identity, video is the medium that seems to have made an irrevocable difference in how we theorize about specificity, for its chameleon-like interface between film, television, computers, telephones, and even architecture seems forever in flux. Video, in short, never seems present only to itself.

As Philippe Dubois's epigraph suggests, therefore, we need to retheorize specificity in this age of multimedia less in terms of ontology, autonomy, and global theory and more in terms of hybridity, dialogism, and intertextuality. Furthermore, we must adopt local, plural theories to analyze the medium under study in its dialectical relation to competing media forms. No medium exists in a vacuum, and thus should be examined within its cultural, social, and political contexts, whose shifting fields of practice reposition and restructure our sense of a medium's specific identity and relative use value. As Pierre Bourdieu has outlined in his *Theory of Practice,* a proper study of cultural production requires a break with a substantialist position "in order to see each element in terms of its relationships with all other elements in a system from which it derives its meaning and function."[4] Because these relationships are dynamic, with various media competing for primacy within shared fields of practice, defining specificity must also be marked by a struggle "often expressed in the conflict between the orthodoxy of established traditions and the heretical challenge of new modes of cultural practice."[5] Thus, as a "new technology" offering an alternative mode of media practice, video has had to assert its identity against the hegemony of its predecessors, such as film and television, whose identities video itself threatens. These rivalries between media forms and practices within the field of cultural production—video versus television, art video versus activist video, Hi-8 video versus Super-8 film, analog video versus digital video—echo the conservative anxieties of Rudolf Arnheim, whose ghost continues to haunt contemporary media studies, which are often unable to negotiate emergent media practices with residual theoretical models that fail to apprehend them.

Home video is one such practice: the bastard of liberal video rhetoric and an orphan of theory, its illegitimate status has inspired, in part, my motivation to interrogate its specificity within the larger field of video practice and to resuscitate its value as an important component of popular culture. Frequently constructed by intellectuals and journalists as the abject "other" against which favored media practices are measured, home video has yet to inspire serious and systematic analysis but is instead cast to the margins, denigrated and dismissed, misunderstood. If currently there's no place *for* home video, one of the goals of this book is to find it a home in media studies.

More specifically, as the title connotes on several levels, there's no place *like* home video, be it a utopian representation of domestic space, a simulation by professional event videographers, or a decontextualized signifier of cinematic narrative. I begin my efforts to situate home video by comparing its practices with those of snapshots and home movies to understand how home video continues a tradition of ideal family representation as it simultaneously modifies conventions established by still photography and motion picture technologies. Then I negotiate home video between the poles of amateur and industrial practice by contrasting it to special event videography, which is both like home video in its appropriation of the sentiments and familial ideologies of the home mode and yet unlike home video in that, as a commodity, it infuses personal experience with instrumental values. Finally, I analyze home video as an imaginary representation, particularly when constructed by and within broadcast television and narrative cinema, whose own specificities have frequently been defined in opposition to it. As networks and studios increasingly simulate video in the home mode, home video has become increasingly homeless—that is, displaced from the actual field of production inhabited by real families and their home consumer equipment.

Thus as we move from home video as ethnographic practice to home video as textual signifier, we must retheorize the category of "medium" itself as a dialectical synthesis of empirical, material technologies and imaginary, discursive constructs. Therefore, in chapter 1, entitled "What Is Video? Mapping Out Models of Medium Specificity," I trace the conceptual models of video and of medium specificity that underline the analysis in the chapters to follow. Beginning with a brief history of video, I survey three paradigmatic models of the medium that have attempted to specify video from the perspective of a predominant aesthetic. First, by an aesthetic of negation, video was first defined against commercial broadcast television: video as anti-TV. Second, by a modernist aesthetic of inherent properties, video was then defined according to its unique essence: video as video. Third, by a postmodern aesthetic of indeterminacy, video is currently being defined paradoxically as indefinable: video as multimedia.

After diagnosing the problems endemic to all three models, whose techno-aesthetic foundations inevitably lead to determinism, prescription, and theoretical anarchy, respectively, I offer several correctives by which to rethink the technological base from which definitions of a specific medium usually begin. I start with the proposition that a medium should be conceived not as pure in and of itself but as a hybrid constituted by heterogeneity structured according to its dominant mode of production and field of practice. Second, I suggest that a medium should be conceived not according to either/or binarisms structuring it synchronically as one thing or another but according to a diachronic dialectic of residual and emergent technologies and practices gradually transforming the medium across a historical continuum. Finally, I reject the technological determinism rampant in medium theory, which confers upon a medium some autonomous and immanent force of inevitable social change. Instead, as a method to understand the relation between technology and social effectivity, I turn to Althusser's theory of overdetermination, which prevents finding simplistic one-to-one correspondences between technological and cultural transformation, and to

Raymond Williams's concept of soft determination, which acknowledges the material constraints of technology without limiting the potential for human imagination and intention to intervene and shape the course of technological development. Finally, I propose that the category of "medium," while grounded in technologies producing a range of aesthetic effects, may better be conceived as a discursive construct not necessarily coterminous with its physical mechanisms. As it moves between a material base and an epiphenomenal idea, our conception of a medium itself mediates the empirical and the imaginary, such that the ways in which we envision a medium's potential use value or represent a medium as a set of immaterial, inherited cultural codes will exert the same degree of effectivity as its substrate and apparatus on how we define its specificity.

Rethinking "medium" as a discourse rather than as a technology therefore requires a methodology suited to analyzing both its cultural and aesthetic effects. For the remainder of chapter 1, I lay out the categorical constituents of the methodology by which I explore home video in its relation to the academy, knowledge, power, social function, cultural hierarchy, and the competing media of television and film. As the overarching link connecting these methodological categories, medium specificity must be understood as a structuring discourse rather than as a basic apparatus or static field of aesthetic effects. Because distinctions among media are culturally constructed, ideological and imaginary more than ontological and empirical, we must reject media definitions that are self-identical, uniform, and predictable by specifying and mapping out the fluctuating historical relations and concrete cultural contexts that shape, but not determine, the ways in which a medium is both used and envisioned.

In keeping with this imperative, chapter 2, entitled "From Reel Families to Families We Choose: Video in the Home Mode," explores the specificity of video within the local practices and discourses of the "home mode," a phrase first employed by anthropologist Richard Chalfen to describe the amateur media practice of idealized family representation. In this chapter, my goals are threefold: (1) to historicize the home mode by analyzing how the introduction of video and changing family dynamics have modified its conventional practices, (2) to theorize the persistence of the home mode across different media technologies and family configurations, and (3) to recuperate home video as a practice with important social value. The prevailing model of the home mode, developed according to the technologies and aesthetic effects of celluloid (still photography and 8 mm film) and applied primarily to snapshots and home movies, requires some revision to understand the material influences of video on home mode methods of production, exhibition, and distribution. At the same time, this model also suffers from an outdated conception of the post–World War II sentimental, bourgeois, nuclear family paradigm on which it has been based and against which it has been criticized. A historical aberration rather than a sociological norm, the nuclear family has increasingly diminished statistically over the last three decades, replaced by alternatives ranging from single parenthood and gay marriage to "families we choose" among relatives, friends, and colleagues. The value systems of these new family forms mark an important difference between home video and the home mode artifacts that precede it, by breaking conventional taboos and broadening the range of content deemed acceptable in snapshots and home movies.

Of course, despite empirical changes in American domestic life, nuclear family ideologies continue to persist in our culture, from sex education and political campaign rhetoric to TV sitcoms and film melodrama. Nonindustrial media practice, too, has not been immune to the attractions of familialism, as the home mode has dominated twentieth-century amateur photography, film, and now video as well. For many scholars of amateur media, the social reproduction of the home mode from generation to generation, despite changes in technology, has proved to be a puzzling phenomenon. Progressive media critics, for whom video continues to represent the potential for radical practice because of its decentralized mobility and widespread access, generally subscribe to a dominant ideology thesis, a top-down explanation that blames corporate intervention, advertising discourses, and training manuals for containing amateur practice to the home mode, thereby eradicating the potential for an avant-garde culture and reaffirming the family values supporting bourgeois hegemony.

In opposition to the ideology thesis, I offer an alternative explanation for the persistence of the home mode based on Bourdieu's concept of habitus, a bottom-up model that argues for the effectivity of ideologies born of, and embedded in, the material practices, historical experiences, and social environments of the individuals and groups who share them, rather than being forced upon them by some external governing agency. I treat home video, therefore, as a practice shaped first by a set of positive intentions and social functions internal to a tradition of family folklore, whose history predates capitalism, and second by the totality of economic and cultural capital available to amateurs of a particular class, region, and historical moment. As opposed to a dominant-ideology thesis, which is generally advocated to support the disparagement of home video as a corrupt or compromised practice, habitus attempts to understand how amateur agency is directed by the concrete structures of historical materialism rather than by the epiphenomenal ideologies of capitalist rhetoric. Having extricated the home mode from the discourses that position it merely as bourgeois social reproduction, I conclude the chapter with a functional taxonomy of home video to recuperate its value as an active mode of media production for representing everyday life, a liminal space to negotiate communal and personal identity, a material articulation of generational continuity, a cognitive construction of home, and a narrative format for the communication of family legends and personal stories.

In chapter 3, entitled "Modes of Distinction: The Home Mode, the Avant-Garde, and Event Videography," I continue my exploration of home video's specificity, situating it within the discourses of amateurism and professionalism, and comparing its practices with those of event videography, a growing industry of professionals hired to record the important rites of passage and celebrations often previously recorded by amateurs themselves. The primary goals of this chapter are twofold: (1) to restore the category of "amateur" as an economic relation in opposition to industrial practice rather than as a generic label for a set of technologies, aesthetic effects, and media texts; and (2) to position the category of "professional" as a social index marking distinctions of symbolic capital within amateur practices and communities in order to rationalize the economic exploitation of laypersons under the guise of expertise.

Beginning with a heuristic sense of "distinction," I survey the theoretical problems that result from the inadvertent conflations of amateurism with technological, aesthetic, and ideological properties, which transpose onto an economic category that should more properly serve as an umbrella term for any nonindustrial media practice a set of preferred characteristics that ultimately set submodes of amateur practice against each other. For example, although both the avant-garde and home modes are forms of amateur practice, Marxist media critics generally write the history of amateur photography as a struggle between its "true," essential, and inherent radicalism and its "false," deformed, and ideological conservatism. As a result, by making the avant-garde mode coterminous with amateurism, the home mode must inevitably be condemned as the avant-garde's betrayal. To locate the origins of this critical confusion, I attempt to unravel the tangled web of associations among avant-garde and home mode practices in the cinema and writings of Maya Deren, Stan Brakhage, and Jonas Mekas, all seminal figures in the history of amateur photographic art, who frequently elided differences between form and function by describing their revolutionary films as "home movies." The point of insisting on the distinction between modalities is not to construct a static or essential taxonomy by which we must judge, say, Mekas's *Lost, Lost, Lost* as *either* in the avant-garde *or* in the home mode, since clearly the film appropriates the conventions of both at various modal moments. Rather, it is to clarify that each modality expresses distinctive social functions and intentions that should not be confused with aesthetics or with genre. We would do injustice to Mekas's masterwork were we to criticize the film as fraught with incompatible contradictions between its reformist call for an avant-garde community of filmmakers and its nostalgic longing for home and family; indeed, the complexity of Mekas's negotiations among his radical and conservative impulses marks the film as a classic.

Typically, discourses about amateur video generally uphold the avant-garde as progressive and denigrate the home mode as reactionary. For this reason, I turn to the second connotation of "distinction" in its positive sense to argue that home video's aesthetic realism and social conservatism fulfill important, if quite different, cultural functions. I perform this recuperation in two ways: (1) by demonstrating that home video's popular aesthetic, which subordinates formal experimentation to the referential documentation of everyday life, reflects the home mode's attempts, during moments of leisure, to affirm a sense of continuity between life and art; and (2) by surveying and debunking the myths of radical video practice that are often advanced to endorse the avant-garde as inherently more revolutionary than the home mode: myths of art and life as coterminous practices, of a democratic collective, and of the mobilizing power of video technology.

Indeed, appeals to the avant-garde as a preferred mode of amateur practice evoke the third and final connotation of "distinction" important to this chapter: the notion that media practices may function as social indexes of cultural taste and symbolic capital. Here, turning from home video's relation to the avant-garde to its relation to event videography, I explore the category of "professionalism" as a discourse sharing, ironically, the avant-garde's goal of demarcating an elitist camp within the larger field of amateur video practice. Certainly, because they are produced for a fee, wedding videos, for example, ostensibly fall within the sphere of industrial practice; but technologically, formally, and ideologically, event videos are frequently indistinguishable from the amateur videos that they simulate. This lack of discernible textual distinction is therefore transposed as a dubious distinction of expertise necessary to sustain the marketing of event videos as commodities greater in value than the home videos produced by "mere" amateurs. By analyzing event videography's pseudoprofessionalism, adoption of "prosumer" equipment, creation of professional video associations, and defamation of amateurs, I demonstrate the methods by which event videographers manufacture cultural distinctions based on imaginary technological and aesthetic distinctions. Finally, by analyzing wedding and memorial videos as examples of late capital's market penetration into spheres of private life formerly immune to commercial exploitation, I outline several of the problematic consequences of event video's commodification of personal experience.

Having upheld taxonomies in chapter 3 in order to preserve for home video an autonomy from the critical hegemony of an avant-garde practice that would seek its demise, I turn in chapter 4, entitled "Family Resemblances: The Home Mode as Chronotope," to a discussion of home video's fuzzy "relative" relation to domestic television, in effect dispensing with taxonomies of difference for those of likeness. By exploring home video and domestic television programming in their dialogic relations, I reject the model of medium specificity constituted by either/or binarisms that have constructed the media of video and television as antagonists. By localizing my analysis within a shared context of domestic values, I seek to avoid the hasty overgeneralizations and prescriptions that seem inevitable when video and television are compared too broadly—itself a problematic endeavor, since both media, designed by their inventors as interdependent forms, cannot be differentiated entirely at more global levels. Through the analysis of various domestic television programs that share family resemblances with home video, perhaps even anticipating its advent through their adoption of home mode conventions, two primary goals of this chapter are (1) to revise the predominant concept of video as inherently antitelevision by illustrating that within specific contexts of production and reception, their history could be written as a pursuit of mutual rather than antithetical goals, and (2) to affirm that, as one connotation of "family resemblances" suggests, the blurring of taxonomies is as important as upholding them in our quest for a dialectical model of medium specificity.

A more literal connotation of "family resemblances" colors another aim of this chapter: an analysis of the intertextual relation among home mode and TV representations of the American family as two processes of signification that construct our popular notions of what a "real family" means, each mediating domestic experience rather than reflecting any determinate social reality of family life. Like home video, domestic television programs adopt naturalized conventions of liveness, presence, and immediacy, which engage the home audience in a parasocial relationship that, while distanced by broadcast technology, links them psychologically to TV families as if part of their social world. Thus domestic television complements home video's idealized representations of how real families might act, especially when audiences look to both media in an effort to recognize themselves on the screen. At the same time, however, both home video and domestic TV, by documenting domestic life as a series of performances recorded for viewing, expose the inherent theatricality of modern families by foregrounding the conventional role-playing of their actors, potentially defamiliarizing the family as a natural social unit. In either case, the proximity of real and TV families both on and in front of the television monitor can be traced from *The Adventures of Ozzie and Harriet,* in which the Nelsons played versions of themselves, to *America's Funniest Home Videos,* a series that suggests domestic television embodies and anticipates a fundamental desire of the home mode previously denied by snapshots and home movies but fulfilled by home video: the opportunity to see oneself and one's family on television.

Thus, in noting the similarities between home video and domestic television as media forms representing home and family, I trace the family resemblances among television programs displaying generic identities that would suggest greater internal differences according to the conventional genre theory. For example, at first glance, the series highlighted in this chapter may appear as a heterogeneous bundle rather than a clear-cut group: *The Adventures of Ozzie and Harriet* (situation comedy), *An American Family* (documentary), *The Wonder Years* (dramedy), *The Real World* (soap opera), and *America's Funniest Home Videos* (game show). If, however, we note that each of these series is constituted, for example, by a self-reflexive focus on familial role-playing, the negotiation of past and present, an emphasis on realism, and a desire to record and observe oneself, such family resemblances may indicate that Bakhtin's notion of "chronotope," concerned with tracing the similar time-space relations among various texts, may be better suited than genre as a category by which to specify not only the more important relations among these series but also their relation to the home mode, which shapes the parameters of this chronotope.

In the fifth and final chapter, entitled "The Video-in-the-Text: A Phenomenology and Narratology of Hybrid Spectatorship," I locate home video in its phenomenological relation to film as a textual signifier of narrative cinema, theorizing its specificity not as an empirical practice but as an imaginary construct. Beginning with a historical survey of media discourses that have positioned film and video as rival systems of representation, I argue instead for a revised model that must take into account cinema's increasing hybridity and focus on films in which home video appears as a representation or simulation within the frame of the narrative's diegesis, including such texts as *Sex, Lies, and Videotape, Family Viewing, My Life, Totally F***ed Up,* and *American Beauty.*

As a dialogic method by which film may define its own specificity in the process of reenvoicing the codes of home video within the "master" codes of cinematic narrative, the "video-in-the-text" functions as a hybrid schema, framing the image from an imaginary point of view whose source

cannot be aligned necessarily with either film or video technologies, but oscillates between and among them. These heteroglossic juxtapositions precipitate a complex phenomenology of spectatorship, as point of view shares the
specificities and cultural codes of home video and those of cinema, its host
medium. For example, qualities that have generally been taboo in Hollywood cinema, such as direct address to the camera and low-resolution
image, de rigueur in home video, have increasingly been borrowed by
mainstream filmic narratives as hybrid codes effecting a limited transformation of classical cinema spectatorship. The divergent dialect of home
video, marked by intimacy, ritual causality, and authenticity, confronts the
monoglossic spectacle, linear plots, and gloss of commercial cinema, setting up a dialogue between two semantic and axiological belief systems,
between worlds in conflict.

To explore the complex of effects precipitated by this video-in-the-
text, I break the chapter down into four sections. First, I define its general
functions and characteristics as an "imaginary apparatus." Second, I map
out a phenomenological methodology by which to apprehend how a hybrid schema influences reception and transcends any technologically determined subjectivity. Third, I perform a narratology of the video-in-the-text
in terms of its transformations of cinematic point of view, identification,
and time-space relations. Finally, I focus on several semantic aspects of the
video-in-the-text, including cognitive mapping, Oedipal conflict and resolution, failures of communication, the impulse toward self-analysis, and
the seductions of vicarious experience.

▶──

A Note on Methodology

As this introduction might indicate, my methodology appropriates a variety of critical paradigms, ranging from Marxist cultural studies to phenomenology, which at times may appear incompatible as grand theories.
Therefore I should clarify at the onset that the aim of this study is neither
to advance a philosophical synthesis nor to establish a governing law of
medium specificity and a definitive taxonomy of home video. Indeed, my
project interrogates the efficacy of all monolithic theories whose residual
binarisms frequently elide or distort emerging hybrid technologies and
practices. As Gilles Deleuze and Félix Guattari have pointed out, global
paradigms, with which they refuse to play "take it or leave it," are themselves constituted by hybridity: "As if every great doctrine were not a *com-
bined formation,* constructed from bits and pieces, various intermingled
codes and flux, partial elements and derivatives, that constitute its very life

or its becoming."[6] Because no single critical perspective may illuminate home video's specificity in its complexity, I propose a series of plural and local theories rather than Theory, if you will, which prevent reducing the medium to any univalent dimension. Instead, like a palimpsest, my discussion elaborates a series of theoretical and critical models that point ahead to and return back on each other, linked by the following overarching concepts around which they cohere.

Context and relation. Rejecting any conception of home video as a self-identical object or idea in and of itself within a hypothetical vacuum, this study insists on locating the medium within a network of mutually related sets of determinations embedded within a specific, local, historical field of practice. In the words of Stuart Hall:

> The meaning of a cultural form and its place or position in the cultural field is *not* inscribed inside its form. . . . The meaning of a cultural symbol is given in part by the social field into which it is incorporated, the practices with which it articulates and is made to resonate. What matters is *not* the intrinsic or historically fixed objects of culture, but the state of play in cultural relations.[7]

Genetic structuralism. This paradigm, succinctly referred to by Bourdieu as the "objectivity of the subjective," combines an analysis of the social and material formations constitutive of any field of practice with a complementary analysis of individual intention and lived experience operating within them, such that the structures constraining practice, rather than being mechanical, predictable, and permanent, may be broken down and rebuilt at different stages in history. As Randal Johnson points out: "Subjectivism fails to grasp the social ground that shapes consciousness, while objectivism does just the opposite, failing to recognize that social reality is to some extent shaped by the conceptions and representations that individuals make of the social world."[8] Thus a dialectical synthesis of both, between the social and individual, structure and intention, material and imaginary, unconscious and conscious, convention and originality, or text and spectator will enlighten and enrich a more holistic analysis of home video practice.

Stratified history. Rather than proposing a definitive taxonomy, this study isolates home video's specificity as a series of synchronic attributes only for the heuristic purposes of analysis, when more accurately the medium has developed along a continuum of diachronic transformations. As Jean-Louis Comolli describes it, a stratified history of media practice is "characterized by discontinuous temporality, which is recursive, dialectical, and not reducible to a single meaning, but rather is made up of types of *signifying practices* whose plural series has neither origin nor end."[9]

By reconceiving the discourse of medium specificity as a dialectical relation among media technologies and cultural practices, we may better understand the dynamics of home video in its particular historical moment—for we may never be sure which of its uses will remain dominant, which of its properties will remain unique, and whether the medium itself will recede as new media emerge to take its place. Indeed, with the invention and diffusion of digital technologies, the term "video" itself must be redefined according to theories as resilient as the medium itself. It is my hope that the models presented here may prove as elastic in accommodating whatever changes in specificity the future may bring.

[1] *What Is Video? Mapping Out Models of Medium Specificity*

In any case, we would agree, I think, that giving up specific identity doesn't hurt anyone, and perhaps it's better that way.—On the contrary, it always hurts a lot, that's the whole problem.
   :: Jacques Derrida, "Videor"

Defining the medium of video has historically been a perplexing project. Written primarily as the evolution of its technology and artifacts, video's history over the last thirty years has been distinguished by rapidly accelerating change. Since the introduction of the Sony Portapak in the 1960s, transformations of the video apparatus, such as the addition of color, high-resolution monitors, and on-line digital editing, have been inextricably allied with transformations of its aesthetic effects. In constructing this history, a crucial dilemma arises: what should we emphasize—technology or aesthetics? Or to pose the question from an alternative perspective, should either of them be emphasized at all?

Marita Sturken, a historian of video art, confronts the problem of a specificity thesis that differentiates video from other media by cataloging its physical mechanisms and formal attributes, both of which supposedly constitute its unique identity:

> The assumption that the aesthetics of video is a direct result of its properties leads us into technologically determinist terrain yet again. Technologies such as television do not simply appear at [a] specific point of history, they arise out of specific desires and ideologies. However, this distinction does not negate the fact that video has a specific phenomenology, which affects our experience of the medium.[1]

Here Sturken diagnoses a fundamental predicament plaguing video scholarship: any pursuit of the medium's specificity by identifying its basic apparatus as the antecedent cause of its aesthetic potential must inevitably run up against the theoretical roadblocks already sojourned by classical film theory. As Sturken implies, a proper history of video must account for the past cultural determinations of the medium's present forms as well as those of the future. Technological determinism and the discourse of inherent properties retard such a history, for they must inescapably fail to pin down contemporary video's manifold interfaces among an ever-changing complex of multimedia.

At the same time, however, Sturken astutely affirms that video does bear a specific identity, which she is unwilling to give up. But rather than equate that identity with either technology or aesthetics alone, she points instead to desire, ideology, and experience in how we perceive, envision, and use the medium. In this sense, "video" refers to an immaterial perception as well as technological object, or more precisely the dialectic of both, for what we know about the medium may be determined as much by imagination as by observation. That is, both hypothetical and empirical, video's specific identity mediates the poles of discourse and technology. It is precisely this dialectical concept of medium specificity that will frame my interrogation of home video in the chapters to follow. We must therefore systematically map out the methodological parameters of this revised model in greater detail.

▶
_____

Techno-aesthetics: Problems of Causality and Ontology

As a once dominant, now residual, but still pervasive paradigm of medium specificity, the techno-aesthetic approach that explains medium effects as determined by technology betrays two problematic tendencies: mechanistic causality and prescriptive ontology. Mechanistic causality tends to explain a medium's social and aesthetic effects as a direct result of its basic apparatus. In turn, prescriptive ontology codifies mechanistic causality as a set of rules dictating practice. Perhaps an example from early television will illustrate their dynamic. As Stuart Marshall has noted, because the small size and low-resolution image quality of prototypical TV monitors rendered the detail in long shots incomprehensible, producers and directors of early television programs were often motivated, but only in part, to reproduce performers' faces on a human scale so that home viewers might see them more clearly. Perceiving these close-ups as determined entirely by video technology, in essence one of its inherent properties, many media critics concluded that TV should therefore represent interpersonal domestic conflict as best suited to the new medium. Marshall points out, however, that because early television experimented with a heritage of entertainment venues to fill its programming schedule, the preponderance of domestic content was more likely indebted to nineteenth-century bourgeois theatrical traditions that reproduce the world in terms of family conflict, rather than being dictated by technological necessity. Thus, he concludes, the close-up is as much inherited by, as inherent to, televisual representation, a convention rather than its determination.[2]

This form of fixed causality diagnosed by Marshall's analysis, better known as "technological determinism," constitutes a transhistorical discourse proposing the belief that media technologies not only dictate aesthetics but organize and govern perception and behavior, acting as the sovereign determinant of social formations and human volition. With Marshall McLuhan, technological determinism found its guru and its peak of popularity in the 1960s as a sound byte: "The medium is the message." McLuhan argued that the forms in which people communicate have an impact, beyond the choice of a specific message, that transforms their culture, personalities, and modes of consciousness.[3] Although rarely cited as an authority by contemporary media scholars, McLuhan's legacy has permeated to a large degree the "commonsense" notions of television in fields such as journalism and congressional politics.

In his effort to stem the tide of technological determinism and debunk its fallacies, Raymond Williams identified and rejected two strains: mechanistic and symptomatic. Mechanistic determinism advances the claim that the media are self-contained, isolated technologies distinct from their cultural environment, yet empowered internally to exercise social effectivity. Symptomatic determinism, more problematic because more common and seemingly complex, adds the corrective that, although communications technologies may be invented as discrete, external phenomena, they inevitably enter into the dominant mode of economic and social production, whose institutions then act on and with the technology to determine its cultural effects. Although symptomatic causality advances on the mechanistic strain by denying technology autonomous effectivity outside of social formations, it disregards the ways in which technologies have been shaped by institutions at their onset, casting the media as a neutral base for a variety of cultural uses, whose positive or negative values will be wholly determined by the good or evil forces that shape them.[4] As we will see in chapter 2, symptomatic determinism underlies the claims of many progressive media critics who argue that the home mode of photographic practice is the direct effect of capitalist advertising discourse, which pushes apparently neutral amateur technologies in the direction toward family representation and away from avant-garde pursuits.

Denying that a medium such as television, for example, can ever be neutral, isolated from its historical relations, Williams maintains that the effectivity of any communications technology must reside also in the degree to which it is itself an effect of the social environment wherein it was produced and of the ideological values that are embedded in its material structures. Thus he explains the invention of television as a cultural form rooted neither in the intrinsic formal properties of its technology (mechanistic causality) nor in the policies of corporate commerce and government regulation (symptomatic causality), but in the much broader social tensions of society at large, in particular, the cultural paradox of "mobile privatization."[5] Although the particulars of Williams's thesis on television are not relevant to this study, what remains important is his insistence on the power of human intention to shape the future of media technologies and institutions, whereas technological determinism "forecloses the possibility of imagining alternatives of any sort, because changes in cultural access and tastes through new technologies threaten the very cultural institutions through which intellectuals and other arbiters of minority culture themselves find power and prestige."[6] Williams's emphasis here on technology as a site of cultural struggle will be elaborated in the third chapter's analysis of media discourses that uphold avant-garde practices as amateur video technology's inherent destiny.

The very idea that video technology would possess a set of intrinsic features returns us to the second constitutive aspect of the techno-aesthetic paradigm, the discourse of "inherent properties," an ahistorical methodology by which a medium is specified according to its supposedly unique, autonomous, formal attributes. As Noël Carroll has observed, this discourse betrays both an internal and external component. The internal component specifies the relation between apparatus and artifact in terms of a domain of legitimate avenues of representation and exploration and identifies the range of effects that accord with the special limitations of the medium in question. The comparative component specifies the relation between one medium and another in terms of canvassing the legitimate effects of both by contrast and holds that there should be no imitation of effects between media. For Carroll, the primary problem of a medium specificity thesis based on the discourse of inherent properties concerns its prescription of practices urging artists to narrow their experimentation to a range of effects that will individuate the medium if it is to be regarded as an autonomous art form.[7]

In Roy Armes's book-length study of video, his effort to distinguish the medium illustrates the slippery slope that links Carroll's internal and external components. At the outset, Armes is careful to note that "video's very versatility and flexibility as a medium repulse any simple attempt to grasp its 'essence' or 'specificity.'"[8] For example, in cataloging video's technological features that produce a sense of immediacy, Armes warns against mistaking these effects as technologically determined: "These are not qualities inherent in tape work, nor are they inevitable outcomes of electro-magnetic recording, but they are crucial factors in how we—as an audience—approach sound and video tapes" (110). However, in contrasting video to film, the comparative component, he claims that "even at its most sophisticated technical level, video remains a personal medium throughout the whole production cycle. This is an immediate difference between video and film" (187). If not a determinist, Armes does let essentialism sneak back into his analysis, because he does not situate video's relation to film within local and historical fields of practice that would blur any so-called immediate differences.

The discourse of inherent properties reflects an ontological idealism, in which our wide-ranging cultural experiences of a medium are collapsed onto one experience of its form, such as "personal," apprehended as identical to the medium itself. Void of their historical relations and cultural contexts, such ontologies lead to three fundamental problems: (1) a reduction of video's heterogeneity to one set of determinations, such as electro-magnetic tape or synchronous sound; (2) a conflation of ontology and history, by which inherited conventions, such as vérité shooting techniques, are reified as intrinsic aspects of the video apparatus; and (3) ontological prescription, which mandates a rigid set of exclusive rules precluding competing media from pursuing overlapping fields of practice in which both may excel.

Having diagnosed in general terms the constitutive elements and theoretical drawbacks of the techno-aesthetic paradigm, we now return to the question "What is video?" to survey how three critical models intended to find a definitive answer according to that paradigm have inevitably led to dead ends. As previously noted, video's diversity of formats, flexibility as a means of recording, and multimedia configurations have escaped ontological categories. This chameleonlike quality, video's ceaseless process of becoming, can in part be traced to the specificity of the historical moment of its advent and cultural diffusion. Developed to complement broadcast technologies, videotape and video recorders were designed initially as accessories to commercial network television, whose own agendas inflected the formal and technical properties of the video medium itself. Thus at its origin, video lacked a secure and distinctive identity. Moreover, as a medium coming of age in the postindustrial era of information, video increasingly serves as the audiovisual interface for interactive computer technologies, resulting in hybrid rather than "pure" video forms. From this perspective, the evolution of a unique video aesthetic can be read as the story of video's simultaneous self-discovery and self-abnegation as a medium, which can be charted in three roughly chronological periods: (1) video's struggle for independence from the specificities of television, (2) its establishment of autonomy as a respectable art form, and (3) its return to indeterminacy as an adjunct technology of multimedia.

▶ _____

Video as Antitelevision: An Aesthetic of Negation

As an accident of history, portable video technologies at first contested the tenets of technological determinism. Coinciding with the publication of McLuhan's deterministic theories of TV, video's first practitioners appropriated the medium precisely in opposition to television. Thus video practice signified antideterminism and the negation of institutional mandates by advocating cultural resistance to, and formal experimentation with, a medium previously theorized as monolithic and inert. Shattering David Sarnoff's broadcast model of television—a single point of production (the professional studio) with a constellation of receivers (domestic television sets)—the invention of the Portapak, video recorder, and videotape provided artists, activists, and home consumers access to a new medium. These constituencies, concerned respectively with art, politics, and entertainment, began to distinguish an inverse relation between "television," regarded as a conservative medium of transmission (of other media forms, of capitalist ideology, of conventional programming), and "video," championed as a revolutionary medium of transformation (of art institutions, of the establishment status quo, of the network schedule).

In general, video artists adopted two strategies of negation: the first, a minimalist aesthetic and assertion of real time to interrogate and detoxify passive television spectatorship; the second, a modification of the codes of commercial television through quotation, allusion, parody, and protest in order to cue recognition of, and distanciation from, the objectionable qualities believed to reside in industrial forms.[9] Perhaps the video installations of Nam June Paik best represent these strategies of deconstruction and demystification: his prepared televisions (*TV Chair*, 1968), minimalist video sculptures (*TV Garden*, 1974–1978), recontextualized monitors (*TV Bra*, 1969), and distortions of received broadcast signals (*Magnet TV*, 1965) stripped television from its institutional and domestic environments to restore video as an aesthetic object. The premier avant-gardist who turned television against itself, Paik rejected the popular conception of the medium as a mass commodity of entertainment and challenged the way in which TV was experienced as an object in daily life.

Social activists who were contemporaries of Paik appropriated video to intervene in the politics as well as the aesthetics of television by organizing alternative production practices (variously referred to as street video, community video, or grassroots video) into modes of communication, consciousness-raising, and social change. Groups such as Videofreex, People's Video Theater, Global Village, and Ant Farm produced, distributed, and exhibited innovative underground documentaries, posed audiovisual culture as a positive alternative to print, and attempted to sublate the art/life dichotomy reified by bourgeois capitalism through integrating video into everyday practice. In 1971 Michael Shamberg, a leader of the Raindance cooperative, advocated the decentralization of television for being an aesthetically bankrupt and commercially corrupt medium in a manifesto entitled *Guerrilla Television*. The appellation has persisted as a generic term for alternative video practices sharing similar sociopolitical goals.[10]

Like early video artists, video activists developed their rhetoric in opposition to commercial television, advocating a populist platform of media literacy and widespread access by which a democratic, collective video practice might counter the demagoguery of corporate and state hegemony. So far, so good. Yet when we look more closely at the particulars of this rhetoric, we realize that its call for a revolutionary media practice is often less informed by efforts to transform the cultural field than it is rooted in a faith in the transforming powers of technology itself. For example, Hans Enzensberger's progressive media theory advocating active two-way feedback over passive one-way reception tended to overestimate the mobilizing power of video, thought to enable each receiver to function as a transmitter as well.[11]

Although Enzensberger never claimed that communications technologies alone could enfranchise democratic discourse within a political system structured by centralized institutions of power, activist rhetoric often enthusiastically simplified his model according to the logistics of mechanistic causality, which bestowed upon the video apparatus the power to catalyze political change. Hailed as the savior of alternative media practice, the Sony Portapak, like a deus ex machina descending on the masses enslaved to the dictates of commercial networks, "released the medium from the economic, ideological, and aesthetic confines of the television studio."[12] True, the appearance of the Portapak marks a convenient date from which to begin writing a history of video art and activism. Still, this rhetoric of liberation overlooks how the Portapak ironically reimported into the discourse of video's specificity an inverse determinism prescribing for video uses limited wholly to those defined against television, a binary model of either/or distinctions in need of deconstruction.

Inevitably, catalogs of formal effects imagined as "proper" to the video apparatus were prescribed. For example, the unobtrusive nature of video's lightweight equipment, its synchronous recording of sound and image, and its capacity to replay information instantaneously—technological properties that may elicit an indefinite range of uses—were geared by activists toward a naturalist approach translating style as "reality" and epistemology as "truth" in a system of representation that measured other formal appropriations of video as deformities against which an essential activist practice must be restored.

Video's aesthetic of negation, its identity as antitelevision, has bequeathed a legacy of binaries structuring contemporary video theory (with video in the first position, TV in the second):

transformation/transmission
heterogeneity/homogeneity
self-determination/determinism
active/passive
producer/consumer
amateur/professional
autotelic/instrumental
raw/cooked
true/false
decentralized/centralized
democratic/totalitarian
two-way/one-way
individual/mass
tactic/strategy
innovation/convention
revolution/oppression

In general, these oppositions polarize video as a medium of self-realization and television (and those video practices that "imitate" TV) as a medium of repressed consciousness. Although this binary model may have served the important political function of rescuing video from absorption by television in the late 1960s, the ongoing transformations of each medium decades later, in particular the breakdown of network monopolies and the rise of the Internet, render it outdated as a method to evaluate contemporary media technologies and practices.

Video as Video: An Aesthetic of Inherent Properties

Once liberated from its dependence on television, to achieve autonomy as a medium standing on equal footing with the traditional arts, video would have to be reintroduced on its own terms, develop its own practices, and argue for the validity of its own aesthetic value.[13] Whereas the discourse of inherent properties surfaced as the implicit consequence of video activism's deterministic rhetoric, video artists explicitly and self-consciously pursued video "as itself" in artifacts foregrounding the materiality of the video apparatus and experimenting with its technological limits. Early works frequently investigated the formal effects of one particular aspect of video's material specificity: for example, instant replay (*Noise,* Benglis, 1972), live feedback (*Locating #2,* Holt, 1972), and electronic dubbing (*Generations,* Bolling, 1972). Video art's first critics championed these experiments precisely for their assertion of the medium's newfound autonomy. Writing in 1973, Bruce Kurtz proclaimed: "The place for criticism to begin is where artists have begun: with an investigation into the inherent characteristics of the medium and their image-making capabilities."[14] Five years later Gregory Battcock would confirm that "if video is truly a new form it must, to a greater or lesser degree, reintroduce on its own terms qualities and principles that are timeless and universal. At the same time it must remain true to the special qualities that differentiate video from other forms."[15] This emphasis on video's unique image-making capabilities represents the medium's claim for respectability by conforming to modernism's insistent demand that any worthy art form should pursue reflexive, medium-specific enterprises and withdraw its field of practice to its own unique sphere, autonomous from competing disciplines and domains contaminated by instrumentality. Ironically, appearing just at the moment when Western capitalist culture was undergoing a postmodern shift, such attempts to construct video as a modernist art form seemed simultaneously enterprising and anachronistic.

While video technologies proliferated in the late 1970s and early 1980s, particularly in hybrid forms sharing the properties of other media, video art in the Greenbergian tradition seemed condemned to an endless quest for a holy grail. As Sean Cubitt has observed, "every pursuit of the medium's specificity uncovers a new impurity, a new relation between video and the adjacent arts."[16] Yet even contemporary media theorists who intend to avoid defining video as an autonomous medium in and of itself may nevertheless find themselves exercising a labored rhetoric that inevitably resorts to equations of either self-identity or negation. For example, in a footnote to an essay distinguishing video from film, John Belton has resorted to tautology: "I try to distinguish between these different usages by employing *video* to refer to the video signal, broadcast and cable television, videotape recorders, and videotapes (and videodiscs) and using the term *video art* to refer to video art."[17] He goes on to refine this self-identical equation by conflating television and the technology of video as

> purely mechanical forms of reproduction. Their value lies primarily in the value of the things they reproduce and only secondarily in the mode of their reproduction, that in many instances is identical to the mode of production that they reproduce. Video art, however, problematizes video as a means of reproduction by calling attention to the medium of video as a medium. Video art makes visible that which is generally kept transparent. In this way it explores the nature of the medium and offers the possibility of a revolutionary way of seeing through it.[18]

Belton appeals to the realist/formalist binaries of epistemological modernism, which structure video either as a blank recording technology transparently transmitting content or as an experimental medium transforming its materials and foregrounding their visibility as art. Ultimately, one logical conclusion of his argument is that video art is self-reflexive television, an extrinsic relation limited to the aesthetic of negation.

This type of approach, although defining video in its comparison to competing art forms rather than in a void, still must beware the tendency to essentialize what ought to be localized as a specific historical relation. For instance, in a seminal article shaping the parameters of video criticism, Rosalind Krauss, after viewing a self-encapsulating performance video by Vito Acconci, proclaimed for the work's self-regard "a narcissism so endemic to works of video that I find myself wanting to generalize it as *the* condition of the entire genre."[19] She describes the specific technological properties only the medium of video could lend (at the time) to an expression of Acconci's narcissism: "Video is capable of recording and transmitting at the same time—producing instant feedback. The body is therefore as it were centered between two machines that are the opening and closing of a parenthesis. The first of these is the camera; the second is the monitor, which reprojects the performer's image with the immediacy of a mirror."[20] Here Krauss transposes her acute diagnosis of narcissism as a metaphor for the identity of the performer onto the identity of the medium in two ways. First, she grounds the term "video" within a specific configuration of the apparatus (live feedback) that she represents as constitutive.[21] Second, she conflates the medium of video with the genre of performance that appropriated it, reversing historical causality by suggesting that the properties of the video apparatus beg for solipsism, when Acconci, having already staged his narcissism in other venues, incorporated video into his repertoire upon realizing how well it could be adapted to suit his art. In short, although Krauss grounds her theory of video's inherent narcissism in a psychological model rather than in the apparatus itself, by equating the artist's intentions with video's dominant aesthetic, her epistemological approach makes it difficult to distinguish if narcissism is the property of the artist or the medium.

Beyond the staging of narcissism, performance artists appropriated video to their craft for a variety of reasons: to work alone, to control and edit the final piece, to repeat a performance without having to be physically present, and to buffer personal revelation from an audience in real time within the safety of an electronic space. I would like therefore to counter Krauss's example with two instances of video practice that also exploit specific properties of the medium yet draw out effects different from narcissism: the autobiographical essay and the confessional. Discussing the works of Ilene Segalove and Lynn Hershman as examples of the "new autobiography" in video, Michael Renov has specified the major characteristic of this trend precisely as antisolipsistic. Rather than fused with itself as in Krauss's description, the subject in these videos is split. Thus the camera and monitor do not inherently bracket out the world and enclose the video artist within electronic parentheses but rather "undertake a double and mutually defining inscription—of history and the self—that refuses the categorical and the totalizing."[22] In these tapes, rather than a set of opposing mirrors, the camera and monitor interact dialectically: while the screen functions as a mirror that reflects back on the subject, the lens acts like an eye that looks out on the world to constitute the subject through and with another rather than alone with itself. These works exploit properties *specific* to video, in that their effects are nearly impossible, say, in film. Obviously, however, these properties are not *inherent* to video, as both the narcissistic and essayistic impulses appropriate the same medium in distinctive ways.

A second example of video practice that illustrates the historical moment of video's specificity is the "video confession" also discussed by Renov, including works such as Maxi Cohen's *Anger* and Wendy Clarke's *One on One* series.[23] Here the subjects are neither narcissists nor essayists, but confessors. They imagine the video apparatus neither as a set of opposing mirrors nor as a roving eye moving outward and inward, but as a two-way glass: a window that delivers the profilmic to an absent gaze and a reflective surface that reintroduces the subject to itself. Renov cautiously disavows any claims for a confessional potential intrinsic to the apparatus; instead he traces the long and varied history of the confession as a practice preceding video by centuries, which in its current therapeutic form during this stage of late capital finds the specificity of video suited to its display and circulation. That is, both the history of their practices as well as their characteristic properties link the medium of video and contemporary modes of confession as vehicles for the transmission of autobiography, the intimacy of face-to-face encounters, and the privacy of an enclosed space. Thus video confessions exploit properties specific to the medium, but defined more appropriately within a discursive and cultural context.

Conceiving of the apparatus variously as mirror, eye, or glass, these works demonstrate that the medium lacks a basal ontology. Video's so-called inherent properties are, rather, revealed to be enabling fictions, which reconfigure the differential aesthetic, cognitive, and cultural effects of individual works into metaphorical models of the apparatus that justify the artist's preferred mode of practice rather than capture video's essence. The fundamental error of the discourse of inherent properties, therefore, is its myth of objectivism, which as Ernest Larsen has concisely put it, fixes "rules as the stereotyped residue of historical relations."[24]

▶
_____

Video as Multimedia: An Aesthetic of Indeterminacy

In the 1970s and 1980s, the modernist pursuit of video's specificity often mistook video's available technologies, received conventions, and institutional dictates for intrinsic aspects of the medium. In the 1990s and beyond, video's manifold affiliations with multimedia, its hybrid relation to cinema, and its protean digital applications have even further attenuated the search for video's inherent properties as a fruitful venture. Increasingly, video serves as an interface networking previously disparate media: the computer (CD-ROM, streaming, downloading), telephone (teleconferencing, electronic mail, chat rooms), photography (photo CD, video printer), television (VCR, TiVo), and architecture (video walls, closed-circuit surveillance). Neither an autonomous medium free of all links to other forms of art and communication nor entirely dependent on any one of them, video can be described only by the plurality of its multimedia relationships rather than as pure in and of itself.

For example, video's supplemental relationship to film has transformed cinematic production and reception. Preproduction crews routinely employ video to record rehearsals, animate storyboards, scout locations, and check for continuity. Technicians transfer film using on-line digital technologies to increase the speed and flexibility of editing and allow for the rapid execution of special effects. And the rise of digital video cameras and projection systems may ultimately replace celluloid technologies altogether in the not too distant future.

Thanks to the ubiquity of the videocassette, DVD, and Internet, audiences more frequently view motion pictures at home than in movie theaters, privatizing a previously public viewing environment. Furthermore, the escalating insertion of video images within filmic narratives has modified the phenomenology of classical Hollywood spectatorship. Appearing in works such as *Death Watch* (1980), *Sex, Lies, and Videotape* (1989), and *American Beauty* (1999), video scan lines, pixels, SLR framing, LED messages, and shuttle effects make up a new set of hybrid narrative codes.

If filmmaking practices have increasingly incorporated video technologies, computer digitization threatens to subsume the distinction between film and video altogether: "There is overwhelming evidence that all-digital media will predominate in the interactive technologies of the future. Being digital, the technology is capable of drawing on sources of data of all types (audio, visual, textual, numeric) and of presenting these disparate strands as a unified package within a single medium."[25] Digitized video replaces traditional image-making tools with a set of algorithmic models that code and decode the analog image as a series of binary symbols unmediated by physical processes, but stored in abstract structures independent of the dispositions and aesthetic qualities of the original video substrate. Digital icons undermine the authority of the video image and distance the artist from the actual process of image creation: whereas analog video's aesthetic has been valued as immediate, literal, and naturalistic, digitized video is more often construed as contrived, synthetic, and analytic. Thus digitization renders obsolete the specific technological and aesthetic differences championed by video activists and modernists alike.

These indeterminacies of multimedia derail the notion that video's essence will inevitably unfold as it evolves progressively toward its self-identity. Such teleology myths have been countered by arguments claiming that because of its radical heterogeneity (capacity for interface and hybridity) and radical homogeneity (capacity for digital simulation), video straddles the modern and postmodern by simultaneously exploding and imploding media boundaries. From these apparently contradictory qualities, according to this logic, the medium's essence can be extrapolated: the specificity of video is precisely that it lacks one.

Still, as this paradox suggests, not all media theorists have given up the search for video's identity, no matter how indeterminate. In her article "Video in Search of a Discourse,"[26] Lili Berko, for example, proposes the hypothesis that the contemporary protean status of video may be temporary, "the object of a discursive tug of war, stuck in between its own ontological specificities which push it to respond to the heteroglossic imperatives of an as yet unknown future and the pull of the hegemonic imperative of broadcast television and cinema's ongoing attempts to recoup video back into their own discursive nets" (289). As opposed to the media of cinema or television, whose identities seem stabilized by convention, video's specificity remains uncertain, as "the liminal status of video in a postmodern world of re-presentation cannot be tied down to a particular discourse" (304).

Looking closely at Berko's language, we may detect, in her stress on "imperatives" and "ontological specificities," the echoes of technological determinism and inherent properties. Yet in a clever rhetorical move, Berko sidesteps these problems by aligning video with postmodernism, in one respect grounding the medium in a particular discourse after all, but in another denying that postmodernism lacks any stable, singular discourse itself. Thus as a "postmedium" succeeding cinema and television, video fits snugly into postmodernism, since both share a "liminal" status in search of their own specificity: "The final position of video within a social form which has not yet been born will finally resolve itself through the discourse which ultimately comes to define it" (305).

Ironically, in the process of liberating video from the hegemony of modernist imperatives, postmodernism recoups the medium in a teleology myth of its own. As Maureen Turim has convincingly argued, such speculations derive from a strand of reflection theory that defines video by a logic that emanates from the culture at large, in particular postmodernism, which "tends to homogenize all issues of culture within the internally contradictory, but nonetheless fixed, present social order."[27] Thus as interchangeable terms of an expressive causality, video reflects a mirror image of postmodernism, which in turn provides for video its essence as its own discourses unfold toward some yet unknown future. In the final analysis, this sort of functionalist analogy comparing video to postmodernism—which oddly determines the medium's ontology by its indeterminacy—must equivocate about the possibility of demarcating a distinctive object or discourse for analysis.

Postmodern media criticism typically propagates two theses about video's specificity, each with its own set of methodological and epistemological problems: (1) that video is atheoretical, a postulate derived by a reversion to modernist categories that fail to apprehend emerging media forms exceeding the familiar categories of canonical high theory; and (2) that video is asocial, a postulate derived by a resignation to a philosophical relativism that voids the requisite critical vocabulary necessary to an adequate analysis of media as a cultural form shaped by specific social relations.

The first methodology, best represented by Fredric Jameson, diagnoses video as symptomatic of postmodernism and thus as congenitally immune to theory. What marks the medium's unique specificity according to Jameson is, again, its indeterminacy, video's unremitting fluctuations that disrupt unities of time and space, interchange positions of subject and object, and instantaneously transform one's critical perspective, thus preventing "the impossible, namely, a theory of video itself—how the thing blocks its own theorization becoming a theory in its own right."[28] Jameson's "theory" of video is compromised by several shortcomings: (1) the logic of symptomatic causality that identifies video practices as uniform expressions of late capital's economic structures,[29] (2) an epistemological conflation of broadcast television and experimental video joined by a misappropriation of Raymond Williams's concept of flow as the only commonly accepted framework of analysis,[30] and (3) a distrust in new, unfamiliar practices and developments outside canonical discourse.[31]

The second methodology, best represented by Jean Baudrillard, posits video as inherently hyperreal, eradicating all sense of social reality. Baudrillard's concept of simulation, derived for the most part from his analysis of electronic media, imagines

> a process of social entropy leading to a collapse of boundaries, including the implosion of meaning in the media and the implosion of media and the social in the masses. The dissemination of media messages and semiurgy saturates the social field, and meaning and messages flatten each other out in a neutralized flow of information, entertainment, advertising, and politics.[32]

Again, like Jameson, Baudrillard misappropriates Williams's concept of flow as a mechanical determinant of cultural processes rather than as one of its effects, by suggesting that video's simulacral and interchangeable codes inevitably confound all sense of the "real" and of history.

What does it mean that decades after its appearance, video remains so elusive that two of our most influential scholars have declared the medium beyond our human capacity to theorize about its identity or to intervene in the course of its cultural use and development? Must we accept video, finally, as an allegory of the fallacy of the specificity thesis altogether? Should we abort our efforts to understand its various causes and effects, content to stand back or afraid to move forward as helpless observers who simply watch video fulfill its inevitable yet inscrutable destiny? I think not. If anything, the era of multimedia should beg rather than dismiss questions of medium specificity, but tendered from alternative paradigms.

If little about video, what this brand of postmodern media analysis should teach us is that either/or binarisms, prescriptive essentialism, and theoretical erasure have circulated a quest for ontology that chases its own tail. Any search for video's self-identity will inevitably lead to tautological definitions by which video mirrors itself in a *mise en abyme* seemingly immune to historical relations and cultural context. This asymmetrical relationship between a residual paradigm in confrontation with emergent media forms precipitates a critical lag, calling for a need to remap models of specificity not only to advance our understanding of video's "nature" but because, more importantly, such models in their most instrumental forms continue to exert tremendous effectivity as institutional and industrial discourses governing media distinctions in the maintenance of power hierarchies. Video continues to be imagined, represented, and marketed as a basic apparatus with essential properties by museum curators, social activists, broadcast journalists, event videographers, equipment manufacturers, television producers, and studio executives, for whom the issue of video's specificity is central to the identity of their own ideological and commercial enterprises—at the expense, to be sure, of those with which they compete. Thus as a site of struggle between a complex field of practices seeking self-affirmation and a rationale for their continued existence in an increasingly competitive economy, video's specificity might better be explored as a fabricated discourse rather than as a technological apparatus, as a medium of cultural as well as aesthetic effects. We turn now to the elements of a new paradigm by which that specificity may be mapped.

▶───

Rethinking Technology: Hybridity, Soft Determination, and the Imaginary

As we begin to remap a model of medium specificity in the direction away from a techno-aesthetic paradigm toward the realm of discourse, we must recognize first that nearly all discursive notions of any medium, such as Krauss's conception of video narcissism, do begin with some idea of its technological base on which more imaginary conceits about its essence are constructed. But rather than identify a medium according to its ontological purity, predetermined effects, and material apparatus—the primary aims of techno-aesthetics—we must instead rethink a medium's technological base as constituted in hybridity, as an effect of social and cultural determinations, and as a set of discursive codes to apprehend its historical rather than essential specificity. Video in its various manifestations—for example, as a multimedia interface, as an invention designed to supplement television, or as a representation simulated by narrative cinema—demonstrates that its apparatus can be defined only within its various relations and contexts rather than be reduced to any universal, basic level.

The lessons of multimedia should therefore teach us not that previously "pure" forms of media, such as film, television, or video, have been "adulterated" but that all audiovisual media have always already been constituted in hybridity, by the fragments of representational practices that precede them:

> The designation of a specifically multimedia form within digital technology at the end of the twentieth century serves as a reminder that combined audiovisual technologies have been in continuous use since the mid-nineteenth century. Victorian dioramas, silent cinema, slidetape presentations, theatre sets, happenings, Disneyworld rides, are all examples of public forms of multimedia.[33]

Rather than trying to pin down a medium to a certain set of elements inherent to it, a more successful approach would map out those elements highlighted at specific historical moments and within specific fields of practice such that their foregrounded preponderance shapes our consciousness of the medium's identity: one that must always be tentative, contemporary, fluid, open to revision, to historical transformation, to shifts in cultural power.

Raymond Williams's concepts of "residual" and "emergent" tendencies may assist us in our method. In *Marxism and Literature,* Williams designates as "residual" those cultural practices formed in the past but still active among the cultural processes of the present; those practices he designates as "emergent" create new meanings, values, and relationships that may or may not significantly transform the residual.[34] Thus, appropriating these ideas, we might better define the material base of any medium as an ongoing dialectic of old and new technologies. Although Williams's emphasis on continuity demonstrates the power of residual practices to shape emergent technologies through their legacy of historical conventions, we must be willing, however, to break from residual theories that polarize (video versus television), conflate (video equals television), distort (home video is a betrayal of amateur video), or fail to apprehend (video is atheoretical or asocial) new technologies and practices.

Therefore, to measure and evaluate both the persistence and termination of residual tendencies within emerging formations, we need also consider the predominant mode of production and field of practice in which the interplay between the residual and emergent is structured, since a shift in one may favor the residual over the emergent (the case of home video, which appropriates new video technologies to supplement a tradition of home mode practice already instated by photography and Super-8 film) or the emergent over the residual (the case of avant-garde video, which appropriates the same technologies, but with the intention to break from domestic use and revolutionize amateur practice). As Williams noted, we can only negotiate emergent and residual practices, finally, within their cultural dominant.

Therefore, by complementing Williams's ideas of emergent and residual with Althusserian structuralism, we can rethink video's heterogeneity not as indeterminate but as structured-in-dominance, its composite elements ordered by hierarchies.[35] The apparatus, its present uses, or the imagination of alternative practices—each being a relatively autonomous yet mutually interactive determination of any medium—may hold sway in the last instance given the dominant mode of production in which they are configured at any historical moment. Thus, for example, we can rethink Krauss's diagnosis of video not as inherently narcissistic but as structured in dominance by solipsistic performance art, the prevailing field of practice for video art at the time she was writing.

As a Marxist, Althusser defined capitalism categorically as the mode of production determining any structure in the last instance. Certainly, in Western cultures, capitalism continues to exert its powerful hegemony on the design and use of communications technologies. Nevertheless, within capitalist modes of production, there do exist more local fields of practice that fail to uniformly share capitalism's values (e.g., drag shows, rave culture, amateur porn). Even home video, which generally does embrace bourgeois obsessions with family and conspicuous consumption, generally rejects capitalism's imperative to sell its home mode and homemade artifacts on the public market. Therefore, for the remainder of this study, I will speak more often about a dominant "field of practice" than of a dominant "mode of production" to prevent slipping into Althusserian functionalism, which tends to critique media technologies as tools for the social reproduction of capitalist ideological state apparatuses (ISAs).

In her study of communications technologies and social intervention, Jennifer Slack dismisses functionalism's symptomatic causality, outlining instead a model of structural causality,[36] which more accurately explains the presence of dominant media practices as well as their transformations. She notes that technology, although a real object, is also part of a changing, complex social formation with specific historical relations that overdetermine its specific practices and effects:

> Assuming that the same physical object exists in a different historically constituted configuration, that same physical object might be a quite different historical object, located in entirely different relationships in the structure in dominance. It is no longer possible to generalize about "the technology" as being identically constituted in all situations or as exercising effectivity based on the expression of its essence."[37]

Slack advocates that media analysts be concerned most with the secondary contradictions existing within a dominant mode of production, for these more specifically clarify how and why a medium is structured.

Perhaps a simple anecdote will illustrate Slack's argument. In 1995 the Academy of Motion Picture Arts and Sciences triggered a flurry of debates within the press by disqualifying the critically acclaimed film *The Last Seduction* from Oscar consideration. Although the motion picture was shot on celluloid with production methods identical to those in the nominated categories, the Academy nevertheless defined the film as television, because it had originally been broadcast on cable networks rather than projected in theaters. By manipulating the various fields of practice within the motion picture industry into a convenient hierarchy reifying exhibition over production as the mark of distinction, the Academy constructed an arbitrary yet powerful media specificity argument that reinforced industry rivalries and championed the economic protection of exhibitors. In this instance, the mode of motion picture production was structured by exhibition methods in dominance. In other contexts, such as independent film festivals, whose board members frequently disqualify entries shot on or transferred to video, it may be structured by apparatus or substrate in dominance. In neither case is the apparatus the crux of the medium's specificity; rather, we see how relations of power mask institutional policy as ontology.

While Slack's model corrects for the twin dangers of determinism and functionalism, to focus on historical relations and cultural context entirely at the expense of technology would swing from one extreme pole to the other. Indeed, although they do fluctuate according to shifts in the dominant order, technologies exert their own important pressures on media practice. We would be misguided, for example, to portray video as an entirely neutral instrument by which any artifact may be produced without regard for the medium's material properties that lend themselves to appropriation: simultaneity of production and reception, recyclable substrate, synchronous sound, extended record and playback modes, portability, clean editing, ease of exhibition, inexpensive cost of duplication, and so on. As Mark Schubin makes clear, specific practices demand the use of either film or video: "A live sporting event must use video, because it's the only medium that can be live," whereas "an atomic blast must be shot with a film camera because the brilliant light will damage a video camera's image sensor."[38] Of course, these "demands" are historically specific effects rather than transhistorical determinants, for improvements in lenses may refine video's sensitivity to strong light in the future. Nevertheless, Schubin's examples illustrate the material constraints of technology on media practice.

To discuss these constraints throughout this study without suggesting the imperatives of technological determinism, I will therefore use the term "determination" as a critical distinction between determinism's closed system of mechanistic causality and a structural description of historical limits. Outlined by Raymond Williams, determination speaks to partial rather than total necessity, in that the materiality of technologies exerts pressures and limits, yet without guarantees.

> It does not predetermine human action in any unilateral sense, but it does make some courses of action more likely than others, if only because it makes some more difficult than others, and also acknowledges that there are, at any one time, certain absolute, often material, limits to the range of human action. Determination also implies that humans learn from their historical experience in ways which create habits and thus inertia, and in ways which provide warnings against certain courses of action and thus make such actions less likely in the future.[39]

Characterized by partial necessity, determination is further distinguished from determinism by the category of "use," which balances material constraints against a diversity of practices that "make do" with media technologies in ways for which they were not intentionally designed. According to Michel de Certeau, "between the person (who uses them) and these products (indexes of the 'order' which is imposed on him), there is a gap of varying proportions opened by the use that he makes of them."[40] For example, human tactical intervention may manipulate or sabotage consumer merchandise, such as Sadie Benning's unexpected appropriation of Fisher Price's "pixelvision" to represent teenage lesbian sexuality, hardly what its marketing executives had in mind when marketing a toy for preschoolers. Of course, although most consumers of media technologies rarely modify them for radical purposes, the important point is that we must analyze a media practice in terms not only of its material constraints but of its potential applications as well. The determinations of human imagination, that is, exert as much pressure on the design and development of communications technologies, opening up certain options while closing others down.

In short, a medium is both a material and a social construct. Whether they are Kodak manufacturers who design home video equipment for commemorating birthday parties or grassroots activists who appropriate that same equipment for documenting illegal working conditions, special interest groups define and give meaning to a medium according to an imagined set of ideological, political, and social goals. As Raymond Williams has persuasively argued, to restore human intention to the development and use of communications technologies differs from symptomatic causality, because intention self-consciously directs its purposes to central social needs and concrete cultural practices: "It is especially a characteristic of the communications systems that *all were foreseen—not in utopian but in technical ways—before the crucial components of the developed systems had been discovered and refined.*"[41] Although here writing about television, Williams echoes André Bazin's foresight in "The Myth of Total Cinema" that the visualization of cinema invariably preceded its discovery and invention: "The way things happened seems to call for a reversal of the historical order of causality, which goes from the economic infrastructure to the ideological superstructure, and for us to consider the basic technical discoveries as fortunate accidents but essentially second in importance to the preconceived ideas of the inventors."[42]

Jean-Louis Baudry has even looked as far back as ancient Greece to speculate that Plato's parable of the cave expresses a seemingly transhistorical wish "to construct a simulation machine capable of offering the subject perceptions which are really representations mistaken for perceptions."[43] Baudry's conjecture, while impossible to prove historically in any empirical fashion, nevertheless implies that a medium is both material and symbolic—a metapsychological projection of primordial urges. Like Metz, Baudry would go on to theorize about cinema spectatorship within a psychoanalytic paradigm that tends to regard desire and apparatus as equivalent. Still, the notion of medium as "imaginary signifier" remains useful if tempered by discourse theory and historical materialism.

Rather than ground "imaginary" within a psychoanalytical theory that would deny the effectivity of conscious intention, in this study the term will refer more broadly to a set of conceptual images circulating as the "common sense" of competing cultural communities whose fields of preferred practice, governing ideologies, and instrumental goals overdetermine a pseudo-ontological model of the medium suited to that community's advantage and cultural investment. A concept of medium as imaginary lies, therefore, somewhere between its material base and an epiphenomenal idea, between its historical specificity and its perceived universal essence. In this sense, discourses about medium specificity are allegorical, circumventing the need to dismiss them purely on the basis of empirical contradiction. Irreducible to its mechanisms and untethered to technological necessity, a medium's specificity, as Stephen Heath has suggested about cinema, may shift from its technological apparatus to a set of metapsychological codes.[44] These codes constitute the imaginary apparatus, the imaginary medium, which despite its relative immateriality may nevertheless generate very real material consequences within a cultural field.

An anecdote about early television may illustrate. According to William Boddy, in the mid-1950s, when live television had been virtually replaced by telefilms as the medium's predominant format, Robert Sarnoff nevertheless presented himself in the *Congressional Record* as an advocate of the aesthetically privileged nationwide dramatic broadcast to justify network monopoly of television against the designs of Hollywood studios.[45] In short, he argued that because TV *could* be live, it should be reserved for broadcasters rather than filmmakers. Thus, by upholding liveness as a quality specific only to television, Sarnoff constructed a model of the medium to his stockholders' advantage. Such a slippery argument illustrates how a single privileged aesthetic property, once intrinsic to a media technology but later extrinsic and outmoded, can gather inertia as an abstracted, autonomous ideology dictating "proper" uses of the medium in defense of an industrial agenda. Indeed, the legendary battles between the motion picture and television industries could be written as the history of competing models of medium specificity.

As Jane Feuer has discussed in her analysis of morning news programs, TV's inherited ideology of liveness in part determines our contemporary perception that recorded programs share a live broadcast's sense of immediacy and intimacy.[46] In more general philosophical terms, George Lakoff supports Feuer's sense of the metaphorical nature of everyday activities:

> What is real for an individual as a member of a culture is a product both of his social reality and of the way in which that shapes his experience of the physical world. Since much of our social reality is understood in metaphorical terms, and since our conception of the physical world is partly metaphorical, metaphor plays a very significant role in determining what is real for us.[47]

Taking a cue from Johnson that our categories of thought are largely metaphorical and thus imaginative by nature, we proceed now from rethinking the technological foundations of medium specificity to mapping out a methodology by which to analyze it as a discourse.

■────────────────────────────────────

Heuristics: Medium as Academic Imaginary

Robert Allen has observed in his anthology on television, "In order to study anything systematically, we must first constitute it as an object of study: as something separable and distinct from its surroundings and foregrounded in our consciousness."[48] Allen reminds us that a "medium," as understood in the academy, is actually a hypothetical point of departure that nominates a schematic exemplar to serve as an objective standard of measure. According to Chris Jenks, this nomination process entails two heuristic impulses: (1) selection, including both our choices of classificatory schemes and our dispositions derived from personal and conventional values; and (2) abstraction, necessary to shift from the particular to the most general level that best serves our theoretical interests.[49] Jenks makes the important point that heuristics, although indispensable, tend to reproduce the researcher's subjective "frame of reference" (a particularly apt phrase for media studies) rather than produce an objective theory. Thus, by locating the specificity of the researcher's position, we will more likely locate the specificity of the medium as a construct.

Along with selection and abstraction, Nelson Goodman adds that we build theoretical objects through processes of "composition" and "decomposition," of dividing wholes into parts, partitioning kinds into subspecies, combining features into complexes, drawing distinctions, and making connections.[50] One trope of this process, metonymy, which hinges on conceptual contiguity, would define a medium by sampling only some of its properties. While a metonymic mapping of specificity provides clarity by highlighting one well-understood or easy-to-perceive aspect of the medium to stand for the whole (such as Roy Armes's upholding of electromagnetic tape to distinguish video from television), we must always be aware that this model might reduce complex phenomena to a uni-level epistemological dimension by excluding other important aspects (video and television share monitors, blurring Armes's distinction). Therefore we need to be cautious when identifying a medium's "basic apparatus," which implies a lack of contradiction between each medium's overlapping relations. If, however, we keep in mind Lakoff's dictum that "basicness in categorization has to do with matters of human psychology: ease of perception, memory, learning, naming, and use," rather than an "objective status external to human beings,"[51] a metonymic mapping of a medium's specificity may assist us in understanding, say, video's phenomenology rather than its ontology.

Metaphor, a second trope of composition and decomposition that hinges on conceptual similarity, would define a medium by projecting familiar patterns from one domain of experience (such as narcissism) onto unfamiliar or new domains of experience (such as video) so that by analogy, an emerging medium may be better understood. Like metonymy, metaphor enables comprehension by highlighting some aspects of the medium while suppressing others, so that its specificity may be referred to, categorized, grouped, and quantified; likewise, it imposes artificial boundaries that may force the complex heterogeneity of a medium within a single epistemological plane (video equals narcissism) or generate tautological circularities (video is postmodernism is video). As Sean Cubitt has observed: "Video is an apparatus, in which the physical machinery and the psychic relations it has with those around it are not just metaphors of one another but to all intents and purposes are models of each other."[52] The same might be said about the history of film theory—Eisenstein's living organism, Bazin's death mask, Baudry's breast, Metz's mirror—more than mere analogies, these hermeneutic figurations hypothesize about cinema's potential to produce meaning. Why stop there? Television as flow, photography as rape, Internet as superhighway: such metaphoric constructions serve as a foundation for the academic imaginary's library of discourses borrowed, returned, and archived for the purpose of circulating and sharing a tradition of ideas, if not scientific truth, about what we know (or think we know) about media. In short, metaphorical models "are 'facts' in the sense that they determine the horizon of our present knowledge; they are not 'facts,' however, in the sense of data supported by empirical evidence and granted eternal validity."[53] If we cautiously heed this dialectic between a need to think analogically and a refusal to think essentially, then metonymic and metaphoric models of medium specificity may serve as useful theoretical signifiers of a medium's cultural, social, and political fields of practice. Perhaps Victor Turner put it best: "There is nothing wrong with metaphors or, *mutatis mutandis,* with models, provided that one is aware of the perils lurking behind their misuse."[54]

Epistemology: Medium as Philosopher

As a generator of hypotheses about practice, a medium is therefore a site of knowledge. Something like a Foucauldian apparatus, the discourse of medium specificity itself mediates a set of communications technologies and its user (or theorist), organizing the two in a relation that reveals as much, if not more, about the user's agenda than it does the essence of the technology itself. That is, in thinking *about* media, we also think *through* them.

> We know that communication media, like language and television, are not simply vehicles for transmitting messages, not simply pipes through which we send messages. If they were, they wouldn't be so powerful. Communication media serve not only as vehicles for transmitting messages; they are used in creating and developing messages—that is, they are used in thought. We have internalized communications media, just as we have internalized language, so that these media can be used not only to transmit messages but as tools to think with.[55]

For example, Gilles Deleuze, in his two-volume work on the cinema, reverses typical approaches to film, neither applying nor constructing a theory to understand the medium.[56] Thinking through the cinema, he is less concerned with a philosophy of film than he is with film as philosophy.

If a medium can be conceived as a structure for ordering our thoughts about the culture at large, we begin to better understand that the appropriations of video by Baudrillard and Jameson serve as metaphors of their global concerns. The important question becomes: At what point does the use of a medium as a metaphor of philosophical speculation reverse itself as an ontological definition of the medium itself? In his discussion of "classificatory schemes," Bourdieu summarizes the problem:

> All knowledge, and in particular all knowledge of the social world, is an act of construction implementing schemes of thought and expression, and that between conditions of existence and practices or representations there intervenes the structuring activity of the agents, who, far from reacting mechanically to mechanical stimulations, respond to the invitations or threats of a world whose meaning they have helped to produce. However, the principle of this structuring activity is not, as an intellectualist and anti-genetic idealism would have it, a system of universal forms and categories but a system of internalized, embodied schemes which, having been constituted in the course of collective history, are acquired in the course of individual history and function in their *practical* state, *for practice* (and not for the sake of pure knowledge).[57]

Here Bourdieu helps us distinguish between media as philosophy and a philosophy of media. The former seeks a Platonic level of abstraction, whereas the latter is grounded in a field of practice. Clearly, if we wish to understand home video, local to the domestic sphere, rooted in a tradition of family photography dating to the nineteenth century, and so widespread that its artifacts can never be collected for global analysis, then the approaches of Jameson and Baudrillard will offer little to this study. My point is not to dismiss their contributions to our understanding of postmodern phenomena but only to cite them as examples of theories about video that are more specifically theories about culture. In sum, as we evaluate the epistemological value of discourses about video's specificity, we must be sure that the medium itself, and not only the philosopher's message, is the primary object of analysis.

■————————————————————————————————

Hegemony: Medium as Power Struggle

As epistemological apparatuses producing knowledge, discourses of medium specificity also seek hegemony. When different communities of users defend their specific practices in a competitive economy wherein survival often depends on penetration into rival fields of practice at one or the other's expense, winning consensus for a predominant cultural meaning of a medium provides one method to control production, distribution, and exhibition to the ideological and financial advantage of one's group. Institutions, both nonprofit and industrial, play the most important role in winning such consensus.

In the early years of video practice, for example, few artists or activists drew rigid boundaries between their respective practices until, as Marita Sturken points out, government granting institutions inaugurated endowments for video; soon afterward, the camaraderie sparked by an enthusiasm for an exciting new medium deteriorated into battles for funding.

> Many of video's funding institutions, such as the New York State Council on the Arts, began to veer away from financing community-based, information-oriented works to funding "video art" by the mid-1970s. This was responsible in part for causing the split in what had been diverse but somewhat coexistent intents among videomakers, in widening the schism that had existed between the two worlds. Irrevocable distinctions were soon made at the institutional level between those who saw video as a social tool and those who saw it as a new art form.[58]

The economic exigencies that first distinguished video art and activist practices were soon transposed and reinstitutionalized as a set of social and aesthetic distinctions masking the more fundamental issue of financial disputation. Soon museums, whose curatorial policies favored formal experimentation and visitor-friendly installations less time consuming than long-playing single-channel tapes, inevitably sanctioned the practice of video art, which more easily accommodated the conventions of museum exhibition. Thus, with more channels of distribution ensuring a public destination for their artifacts, video artists more readily acquired funding, defending their existence and cultural value at the expense of video activists. The hegemonic practice of video as modernist art, of foregrounding its formal attributes rather than exploiting its capacity for social realism, should therefore be located within institutional discourses of power, not within the properties of the medium itself.

■───

Ideology: Medium as Social Function

As illustrated by Sarnoff's defense of live television at the expense of cinema despite his network's dependency on telefilms, the hegemony among media discourses hinges more on ideology than technology. That is, all conceptions of medium specificity structure ideologies already present in cultural formations by organizing practice according to the social functions intended by various user communities. For example, in chapter 2, I argue that within the field of amateur media practice, currently circulating avant-garde and familial ideologies find different expression through the medium of video, resulting in a diversity of forms, none of which should be evaluated as more ideological than another. Because "ideology" and "social function" serve as important theoretical terms throughout this study, more precise definitions are called for.

"Ideology" will refer to a system of representations endowed with a historical existence and social function within a given cultural formation, each system exerting its own logic of development and autonomous effectivity. Thus ideology is neither coterminous with bourgeois values nor reducible to false consciousness, since it constructs subjectivity through the historical conventions of specific material practices whose plural systems of values may either sustain or subvert capitalist modes of production. Although systems of representation will always be subject to the hegemony of predominant modes of production, no field of practice will ever be "liberated" from ideology; instead, subordinate ideologies struggle to win consensus by naturalizing their own historical conventions as more "authentic." Thus we should evaluate a specific media practice not as deformed by, or free of, ideology but as in competition with other practices organized by different systems of representation.

Furthermore, while constituents of every social totality, ideologies neither reflect nor express that totality. That is, cultural practices do not uniformly reproduce a "dominant ideology," since any cultural formation is overdetermined by contradiction. For example, vulgar functionalists can argue that the nuclear family is designed to prop up the prevailing socioeconomic system, but only if they disregard the empirical evidence of rising divorce rates, single-parent families, and gay lifestyles, whose value systems are in tension with capitalist structures of production. While certain practices, such as ritual behavior, may unconsciously reproduce dominant ideology, other practices may be self-consciously chosen to fulfill a variety of cultural functions.

For example, a bride and groom may record home video of their traditional wedding ceremony as tacit confirmation of conventional gender relations, whereas a gay couple in drag may record their commitment ceremony to subvert those very conventions. And as husband and wife, the married couple themselves may on certain occasions record, say, their newborn's ritual baptism, but on others their landlord's shoddy repair work for investigation by HUD, a vaudeville skit for audition to the local talent show, or sex games with their swinging next-door neighbors. By restoring intention within a field of media practice, particularly home video, which

is too often regarded monolithically as feckless bourgeois domestic reproduction, I reject arguments defining fields of practice as uniformly or uni-
dimensionally ideological, choosing instead to describe more accurately the
several modes individuals may actively choose within the field, rather than
being always already passively interpellated or "duped" by the dominant.
In this study, therefore, "social function" will reject "functionalism" to
refer instead to the diverse sets of intentions, desires, behaviors, and goals
of a community (more heterogeneous than homogeneous) sharing ideological systems of representation, which neither determine nor guarantee the
social reproduction of a dominant cultural formation's ideologies, modes
of production, or state apparatuses.

Sociology: Medium as Cultural Distinction

If media practices intend multivarious social functions, they also legitimate
invidious social distinctions. In other words, a medium's defining heuristic
class may itself signify social class power and privilege. In his study of cultural practice, including photography, Bourdieu has argued,

> What individuals and groups invest in the particular meaning they give to
> common classificatory systems by the use they make of them is infinitely more
> than their "interest" in the usual sense of the term; it is their whole social
> being, everything which defines their own idea of themselves, the primordial,
> tacit contract whereby they define "us" as opposed to "them," "other people,"
> and which is the basis of the exclusions ("not for the likes of us") and inclu-
> sions they perform among the characteristics produced by the common classi-
> ficatory system.[59]

If Derrida's method of deconstruction unraveled the hierarchical relations of superior/inferior inhering within metaphysical binaries, here
Bourdieu's theory of practice grounds classificatory schemes within the cultural field to map their transverse relations of inclusion/exclusion, which
structure media practices in oppositions such as avant-garde and kitsch,
amateur and professional, high and low. In a sense, Bourdieu's sociology
repoliticizes poststructuralism by rooting the origin of binaries within the
context of historical materialism (the move from linguistic *différance* to
cultural *distinction*). Furthermore, Bourdieu's notion of distinction supplements Gramsci's concept of hegemony by identifying power struggles not
only between the dominant and subordinate factions within a broad social
formation but also between factions competing in subordinate fields of
practice themselves, thereby challenging arguments for the inherent subversive capacity of subcultures through coalition.

For example, within the field of video practice subordinate to the hegemony of industrial television, avant-garde video and event video vie for respectability through methods at home video's expense: the former through discourses of radical activism and formalism opposed to conservative familialism and realism, the latter through discourses of professionalism opposed to amateurism. Without background knowledge about the contexts in which avant-garde, event, and home videos have been produced, we may be hard pressed to discern among them any technological or aesthetic distinctions: professional event videographers routinely exchange tapes recorded and edited on home consumer equipment for thousands of dollars, while nameless amateurs fail to find (or even seek) museum exhibition for their off-the-wall home video diaries. Regardless, invested groups, such as museum curators and wedding photographers, tend to naturalize avant-garde and professional entitlements as innate dispositions, internal to both practitioner and artifact.

Bourdieu, on the other hand, traces production histories back to external determinants such as income and property (economic capital), education (cultural capital), and imaginary prestige (symbolic capital). Substituting the authoritative discourse of "genius" with the material discourse of habitus, Bourdieu's sociology seeks to understand the cultural histories of social groups that deeply structure individual tendencies toward, for instance, the home mode's bourgeois realism or the avant-garde mode's disinterested formalism. Particularly important to chapter 2, the concept of habitus will help explain the persistent inertia of the home mode over generations of media practice, counter the dominant ideology thesis that, by eliminating capitalist discourses of advertising and instruction, amateur video will be liberated from domestic concerns, and debunk the alternative technology thesis that revolutions in technology alone may revolutionize practice.

___

Dialogism: Medium as Dialect

A medium may be represented as a discursive construct not merely in written language: its specificity may also be simulated by the audiovisual language of rival media systems. That is, one medium, such as cinema, exploits its own system of representation to re-present the system of representation of another medium, such as home video, as a narrative image, a semantic utterance, or an alternative dialect of expression. For example, filmmaker Atom Egoyan has consistently interrogated his cinematic practice by inserting home video sequences within his narrative diegeses. In *Exotica,* when the protagonist mourns the murder of his young daughter, flashes of her smiling image on home video substitute for his more painful, repressed memories. In *Family Viewing*, home video thematizes Oedipal rivalries, as father and son assert their autonomy within the household by stealing, erasing, or taping over old cassettes of their everyday domestic life. And in *Calendar*, whose protagonist photographs Armenian churches for a commercial Canadian calendar, home video supplements his still camera as a tool to colonize, collect, and export his purely acquisitive experience of another culture.

In each film, home video functions as an imaginary medium. Its appearance within the frame elicits viewers' memories and experiences of their own home video production and reception practices, transposed onto and potentially transforming cinema spectatorship. And because the medium's codes may be simulated by professional film and digital equipment, "home video" acts as a cinematic signifier untethered to a video apparatus altogether. Nevertheless, by our capacity to identify "home video sequences" within a narrative film produced in a studio, we acknowledge the history, cultural connotations, and phenomenology of the medium's specificity.

To analyze the self-conscious hybridity of media forms representing each other in and through each other, Bakhtin's theory of dialogism offers a model based on their mutual and interactive relations. In particular, his concepts of reenvoicement and parodic stylization may be modified from their linguistic applications to study the aesthetic interpenetrations, social recontextualizations, and discursive collisions between individualized media systems with distinct intentions and points of view on the world. Informed by Gramsci's theory of hegemony, dialogism also uncovers the political relations encoded in hybrid artifacts, in particular the efforts by the representing or "master" medium to subordinate the represented or "subaltern" medium as "other" to its hegemonic norm. Thus, for example, in chapter 5, I will pursue home video's specificity as it is defined by cinema, which inversely defines itself against home video by opposition.

What Is Medium Specificity?

Although no governing law or global theory can ever completely contain or define the essence of communications technologies, discourses of medium specificity remain essential analytical tools for charting and navigating the plural fields of multimedia practices proliferating at the end of the twentieth century. As our brief history of video aesthetics demonstrates, the techno-aesthetic paradigm that defines a medium by its available apparatuses and internal properties must be replaced by a historically material model emphasizing a dialectic of technology and practice mediated by human desire and intention. That is, at a particular time and in a local situation, a medium's specificity should be defined as those properties that lend themselves to the specific social functions and cultural agendas of invested individuals and groups. In short, a medium's specificity is the historical, technological, and imaginative structuration of its use.

As a structuration, medium specificity must be understood primarily as discourse, not as ontology. Distinctions among media are culturally constructed, and irreducible to the empirical. Techno-aesthetic definitions lead inevitably to determinist and essentialist prescriptions, which imply that a medium is self-identical, uniform, predictable, and predictive. As Raymond Williams has argued:

> The actual production process is a complex of material properties; of processes of signification within these; of social relations between producers and between producers and audiences; and then the inherent and consequent selection of content. To reduce this complex to the "medium," with supposed objective properties governing all these processes and relations is strictly a fetishism.[60]

As a discursive framework, medium specificity can therefore be broken down into methodological models apprehending its various structures of cultural effectivity: a heuristic construct necessary for description, analysis, theory, and criticism; an epistemological machine producing knowledge about the user's or theorist's agenda; a hegemonic power struggle territorializing fields of preferred practice; an ideological organizer of social functions; a marker of cultural distinctions and class hierarchies; and an imaginary apparatus transcending its material substrate as a coded system of representation. Unifying these methodological models is the necessary sense of "historicity," of the diachronic development over time of media artifacts themselves, and our own historic positions at a particular cultural moment that in part determine the very questions we imagine asking as media scholars. Therefore a proper media history can only accumulate models of specificity as a set of "successive synchronies," some of whose times will pass, and some of whose are yet to come.

▶

Conclusion: What Is Video?

What indeed? Perhaps in the chapters to follow we may find an answer. But perhaps the point is not the answer but the question itself. Perhaps the goal is not to define the ontology of video but to analyze the epistemology of ontology itself. Video's identity as a medium remains not in its being but in its use: how it is imagined, designed, marketed, practiced, and represented, rather than what it is. Heterogeneous, a by-product of other technologies, the source of multiple practices, struggles, and contradictions, "video" is too fragmentary a medium for any single discourse to explain its fundamental essence. Instead we must turn to a discourse of pluralities, from video medium to media, from video ontology to ontologies, from video culture to cultures. We must descend from the empyrean of high theory to the ground of case study, from metaphysics to the cultural field, from a question of medium specificity to questions of medium specificities. Keeping in this spirit, we now move from video in general to video in the home mode.

[2] *From Reel Families to Families We Choose: Video in the Home Mode*

During the mid-1980s, at the moment when amateur cinema had emerged within media studies as a field worthy of critical analysis and Super-8 film-making had spawned a lucrative trade magazine industry, academics and journalists invested in home movies soon bewailed the rapid diffusion of video technologies, which had virtually supplanted celluloid within a decade as the dominant amateur motion picture medium.[1] Even video enthusiasts tended to look back fondly on Super-8 with a resigned sense of nostalgia:

> In the home video market, film has also fallen on hard times. The video cam-corder has totally replaced the super eight movie camera in American homes. Baby's first bath or the family backyard barbecue are now preserved on video-tape not film. Even old home movies that had been previously recorded on film are now commonly transferred to videotape for purposes of preservation, or easier presentation on the television set in the family living room.[2]

While home video seemed like a harmless technological extension of the home mode, many defenders of home movies perceived its simplified methods of production and exhibition as somehow too easy and therefore art-less, disposable, corrupt:

> When the video camcorder came along, it seduced the nation with its acces-sibility: no need to change film cartridges every three minutes, no hassling with a temperamental projector, no screen. But folks who switched to video weren't stepping up, they were stepping over, trading away the look and theatricality of film because they shared the same sin that leads to fast-food joints: the I-want-it-now-syndrome. They traded taste for prompt delivery.[3]

Here, this critic transposes his nostalgia into a distinction of taste. Home video not only usurps home movies but signifies a break from the "good old days" and portends a clichéd image of the future: a culture of junkies enslaved to the present moment, seduced by video's simulacral thrall.

Camcorders erase home-movie history as a technology too intrusive and too aesthetically complicated. Its images endure too permanently to have any use value at all in the ephemera of the end of the twentieth century. The floating signifier of Jean Baudrillard, the signifier that can be attached to anything, finds its technological articulation in home video, where family history can evaporate with the push of the button and a different, happier history can be encoded.[4]

Moving from nostalgia to distaste to despair, these snippets only briefly survey the discourses on home video that portray the medium as a menace: to an outdated technology, to a favored practice, and to history itself. Like Arnheim, self-righteous about "new technologies" that perturb their theories about a cinematic practice dear to them, many critics of home video have prejudged its practices, which seem alien to their conventional ideas about amateur film. In particular, for many writers who have published widely on home movies, home video threatens the relevance, longevity, and authority of their discourse. Yet rather than modify or update their extensive research to accommodate home video, many home movie scholars may choose instead to other the new medium as essentially beyond the reach of their critical models, retreating to the polarizing either/or binaries so prevalent in specificity arguments about video.

Although most home movie scholars readily acknowledge that the "home" in "home video" connects its practices to the ongoing conventions of family photography, just as many insist that the "video" marks a definitive break (whose absolute value depends on whether the critic either favors or condemns domestic photography, which video then either corrupts or revolutionizes). Betraying the mechanistic causality of technological determinism, these arguments place greater faith in the power of technology to transform practice than in the structures of the cultural field to regulate use. Thus one goal of this chapter will be to renegotiate the relation between home movies and home video as a historical dialectic rather than as an essential opposition.

A failure to apprehend, accommodate, or appreciate home video often stems not from the essence of the medium but from the historical position of the paradigm home movie scholars have constructed to describe and evaluate family photography. The most influential theorists of the home mode, Richard Chalfen and Patricia Zimmermann, extrapolated their critical models by focusing on practices prevailing in the first few decades following the end of World War II. At this historical moment, amateur media were structured in dominance by technological and social configurations that no longer hold sway: celluloid and the nuclear family. Chalfen, for instance, cataloged the aesthetic effects of the home mode based primarily on his reading of snapshots and home movies, from which he interpreted cultural meanings as essential to the mode rather than as technologically contingent. Zimmermann, in advancing her thesis that the hegemony of bourgeois capitalist values contains all amateur practice to the domestic sphere, based her argument on an ideology of familialism local to the sentimental model of the nuclear family reigning only from the late forties to the early sixties. Although contemporary amateur media practices in the eighties and nineties have been structured in dominance by video and the breakdown of the nuclear family, because Chalfen's and Zimmermann's scholarship broke significant ground, their models have frequently been imported intact to evaluate home video. Inevitably, this application of a residual model to an emerging practice precipitates critical oversights and distortions.

Therefore the fundamental aim of this chapter will be to redefine a domestic model of amateur media practice that can accommodate video and changing family dynamics; understand and explain home video's divergence from, and continuity with, the practices of snapshot photography and home movies; and recuperate the home mode in all its mediated forms from critical discourses that discount its significant and valuable ritual functions. I begin by surveying the basic aesthetic and ideological constituents of the home mode outlined by Chalfen and Zimmermann, noting their individual contributions, but also the limitations that compromise their analyses of home video. I then proceed to historicize their models by analyzing the different effects video technologies and new family configurations introduce into the home mode in an effort to loosen the rigid critical boundaries that apprehend contemporary home video practice as either an insignificant variation of home movies or their illegitimate offspring.

Moving from an examination of the technological and familial characteristics that empirically distinguish the era of home video from the era of home movies, I then interrogate the ideological characteristics that link them. That is, if video and alternative families differ considerably from celluloid and the nuclear model, why has the home mode persisted as the predominant form of amateur media practice? The ideology thesis would blame dominant bourgeois ideology disseminated through advertisements and training manuals for coercing amateurs away from radical or activist projects that might threaten the hegemony of state and corporate agencies. Although admirable for its progressive rhetoric, this argument is founded on problematic arguments regarding false ideology.

Instead I offer up Pierre Bourdieu's concept of habitus, informed by his sociology of practice, that helps us understand that the effects of amateur media are determined more by values and habits already deeply structured in the socioeconomic field than by epiphenomenal ideas circulating in commercial discourses. By grounding the ideologies of the home mode in their material and historical origins, including a tradition of family folklore predating capitalism, I reject polemics that uniformly position home video in essence as a tool for bourgeois social reproduction. Understanding the home mode as a changing expression of culture rather than a static reflection of false consciousness, I take a more celebratory approach to its practices and artifacts to conclude the chapter by reaffirming the values of the home mode's ritual functions.

▶

Defining the Home Mode: From Chalfen to Zimmermann

Because the phrase "home video" frequently serves as an umbrella term for a heterogeneous set of goods, services, and practices (such as consumer video equipment, films on videocassette, and amateur videography), I need first to clarify that future references to "home video" throughout this study will refer more specifically to the amateur practice of video in the home mode. What exactly is the "home mode"? Richard Chalfen, a visual anthropologist whose ethnographic studies of domestic photography generated a series of articles eventually synthesized in book-length form as *Snapshot Versions of Life,* coined the term to categorize amateur representations of family popularized by widespread access to automatic technologies after World War II, in particular snapshot and home movie cameras.[5] Conceiving the home mode as an autonomous field of practice worthy of internal analysis, Chalfen advocated that its unschooled characteristics, personal themes, and autotelic functions be studied distinctly from the professional formal codes, public display, and commercial exchange value of industrial media artifacts. Thus redeeming the value of faded photo albums, boxes of one-reelers, and projectors gathering dust in the closets and basements of middle America, Chalfen opened new doors to reexamine the forgotten archives of amateur images, whose historical importance for interpreting the twentieth century's American ideal might rival the hegemony of Hollywood cinema and network television.

Simply put, Chalfen's model defines the home mode as a pattern of interpersonal and small-group visual communication centered on everyday domestic life. He delimits its practitioners (photographer, subjects, viewers) to immediate family members, relatives, friends, and neighbors (never strangers or mass audiences). In general, its artifacts portray special events, holidays, and vacations as a stimulus for new image making and image review. In particular, its governing themes celebrate rites of passage from birth to old age: monumental events (birth, wedding), new physiological changes (first tooth, first word, first step), new spiritual identity (bar mitzvah, first communion), new social status (boy scout, cheerleader, GI), moments of accomplishment (graduation and retirement ceremonies), landmark accumulation of goods (new car, new house), and, of course, gift giving, which marks the rites of passage with both spiritual and material signifiers of success. More than just souvenirs of the past, home mode artifacts provide families with important cultural functions as well: retention of detailed memories of people, places, and events; transmission of personal stories reformulated into the context of an individual lifetime; kinship affiliation and generational continuity; and connection to the land and accumulated goods.

Chalfen's catalog of characteristics, themes, and functions remains useful for outlining, with important modifications, the general parameters of home mode practices today, including home video. His methodology, however, is jeopardized by formalist fallacies. Chalfen extrapolates his model of the home mode by reading off the internal visible attributes of its artifacts as structural signifiers of everyday life, as statements about how families see themselves in their social context: "Pictures must be understood as evidence of . . . how people have constructed a particular view of the world."[6] This approach conflates general formal conventions with specific cultural intentions, reducing the diverse subjective aims of home mode practitioners to the researcher's interpretation of an objective image. That is, Chalfen assumes that amateurs see themselves in everyday life the way he interprets their appearance in snapshots and home movies. As a variant of reflection theory, this methodology, therefore, cannot adequately account for or explain the broader range of family dynamics and ideologies of home that escape the lens of a Polaroid or Brownie camera.

Concerned more with home mode artifacts as sociological data that increase our understanding of family life in the abstract and less with their history as end products of specific media apparatuses, Chalfen discusses both snapshots and home movies as more or less equivalent, neglecting the significant determinations—constraints and pressures—that still and motion picture technologies may bear upon their intentions and effects, both cultural and aesthetic. Such inattention to medium specificity inevitably prevents him from perceiving potential changes in the form and content of home mode artifacts upon the arrival of video, such as the introduction of sound and extended record modes, which increase the volume of quotidian signifiers available to the home mode practitioner for expression and to the home mode researcher for interpretation.

In *Reel Families*, Patricia Zimmermann seeks to amend Chalfen's ahistoricism. She critiques Chalfen's presupposition that the home mode has been a self-regulating, self-identical practice consistently and uniformly resisting the industrial discourses that have pressured amateur practice to conform to standardized codes of production. Thus she defines the home mode by its external relations rather than by its internal essence. Her contextualist approach locates the home mode in its set of affiliations with the ideologies and economic structures of dominant media practice, in particular Kodak's marketing strategies and professional training manuals, which determine patterns of production and consumption by confining amateur photographic technologies and their artifacts to the domestic spaces of the bourgeois home. Thus, rejecting a static model of the home mode as a transhistorical amateur practice, she redefines it as one effect of changing social and power relations that shape the contours of legitimate practice and preserve professional privilege. In particular, she indicts an ideology of "familialism," which transfers the idea of the integrated nuclear family unit onto all available cultural and aesthetic formations, for inducting amateur technologies in the service of bourgeois social reproduction and for betraying their potential for more radical practices. In short, whereas Chalfen grants the home mode a relatively autonomous status, Zimmermann repudiates it as contingent and ultimately objectionable.

By restoring the categories of context, relation, and history to a theory of the home mode, Zimmermann makes important advances in methodology. Yet in attempting to account for the home mode, her model deforms as often as it explains amateur practices. For example, though Zimmermann rightly argues that a mode of media practice can only be understood as generated by its cultural context rather than as extant in a formal vacuum, she shifts her focus to a purely ideological context whose connections to actual material practice may be tenuous. Tracing the origins of familial ideology not to the social formations preceding the advent of amateur technologies but to advertising discourses succeeding them, Zimmermann's model of historical causality overlooks centuries of cultural practices that had shaped the contours of the home mode by the mid–twentieth century. Whereas Chalfen's model implies that families appropriate amateur photographic technologies to express intentions and desires for kinship and community already embedded in the culture, Zimmermann's model represents these functions as ideological constructs expressly designed to train amateurs to practice only family photography. Where Zimmermann might locate the home mode within the material context of its social functions, instead she consigns it to the epiphenomenal context of false ideology.

While rightly contending that amateur practices must be defined in their diachronically changing relations against industrial practices rather than as synchronically identical to themselves, Zimmermann's methodology often immobilizes these relations in two ways. First, where she might argue for the potential of amateurs to challenge dominant practice with radical alternatives, she implies instead that oppositional tactics are the inherent properties of amateur technologies themselves, thus reducing a truly dialectical relation to an either/or binary by which an amateur practice may be measured as either a pure expression of its technology's radical destiny or its corrupted domestication by familial ideology. Second, by aligning familial ideology with industrial capitalism and thus with dominant practice, Zimmermann must unequivocally reject the home mode as the betrayal of amateur practice. This dichotomy not only obfuscates the category of amateurism but holds the home mode accountable to cultural functions beyond the domain of its ritual intentions.

Finally, like Chalfen, Zimmermann discloses her own brand of ahistoricism. Deriving an ideology of familialism from the sentimental model of the bourgeois nuclear family of the 1950s without acknowledging that its predominance during the immediate postwar era was more a historical aberration than a cultural norm, Zimmermann goes on to argue that by the early sixties, all amateur technologies had been completely co-opted by familialism, thus conflating amateurism and the home mode as coterminous fields of practice. Without providing a history of amateur practice during the years that intervene, Zimmermann transports her model into the eighties and nineties to critique contemporary home video practices as if the same monolithic familial ideologies have been preserved intact three decades later. Having critiqued Chalfen for his transhistorical formalism, Zimmermann tenders a transhistorical ideologism.

In summary, although their models have pioneered the contemporary study of the home mode, both Chalfen and Zimmermann have theoretically privileged the historical moment of their research, transposing the technological and ideological effects of snapshots and home movies onto home video practices.[7] We need, therefore, to revise a model of the home mode that accounts for the specific determinations of home video, including its apparatuses that alter practices of production and reception, and changing ideologies of home that reconstitute the fractured nuclear family as a discursive domestic space. Let us turn first to issues of technological determination.

▶

Home Mode Technologies: From Celluloid to Video

Although Chalfen's focus on the social functions of the home mode prevents his model from slipping into technologically determinist rhetoric, by privileging intention almost entirely at the expense of technology, he discounts the important distinctions between celluloid and video, and their respective material determinations that lend themselves to diverse, if not discrete, uses. He tends to view video technology as just another photographic tool that home mode practitioners take up as updated versions of home movies:

> We have seen that home mode camera users do not take advantage of tech-
> nological potential on any regular basis. While the potential for creating an
> alternative scheme of settings, topics, participants, and conventions exists,
> the possibilities are generally not exploited. Thus, even with the introduction
> of video, traditional patterns of appropriate subject matter and view of the
> world may continue to remain relatively stable.[8]

Chalfen bases this conclusion on his discovery that home moviemakers seemed to replicate the communication patterns of snapshooters, despite the addition of motion and, in some cases, sound. He imagines that video will reproduce the same production and reception patterns in its turn, without considering the significant differences between videography and photography. Although the celluloid substrate does link snapshots and home movies closely, the electromagnetic tape or digital code of video may distinguish it as subject to a different set of limitations and potentials.

Curiously, Chalfen more readily perceives that amateur and professional equipment produce different representations, quoting film scholar James Potts to make his point: "Technology is not value-free: to some extent different technologies dictate the way in which we see the world, the way we record and interpret 'reality,' and they influence the types of codes we use to communicate a message."[9] One might imagine, therefore, that Chalfen would be more sensitive than he is to video's values and codes that inflect home mode communication. True, snapshot photography, home movies, and home video do share fundamental ritual functions that link them within the home mode as a coherent field of practice. Yet we must avoid eliding their specific practices and artifacts, which can be distinguished effectively by differences of substrate and apparatus.

Let us, for instance, compare some of the technological determinations differentiating home movies and home video.[10] The home movie apparatus consists generally of a camera lacking sound recording, projector, screen, lights for indoor photography, and reels of film limited to three minutes of shooting. Its celluloid substrate is relatively costly per foot exposed, cannot be recycled, and requires high light levels and lab processing for proper exposure and printing. The home video apparatus, in contrast, consists generally of a camcorder with synchronous sound recording, VCR, domestic television monitor, optional lights, and videocassettes allowing up to eight hours of shooting. Its electromagnetic substrate is relatively inexpensive per foot recorded, may be recycled, operates in low light levels, and does not require lab processing. Even if, as Chalfen suggests, most home mode practitioners do not exploit the more advanced and complex options that either of these home systems offer (in particular postproduction editing), these basic differences of operation will precipitate differences of production and reception, which in turn may extend home video's range of content and space for interpretation beyond the limitations of home movies.

For example, home movie production methods emphasize brevity, control, and selection. The limited number of feet per reel and the impossibility of reusing celluloid pressured photographers to restrict their selection of material and to shoot only one take. Thus, rather than expose random moments from everyday life, which would require a much greater financial investment, home moviemakers generally film only the highlighted moments of ritual events wherein participants could be posed and conventions controlled in advance of shooting. As well, the absence of sound recording on most Super-8 cameras minimized the narrative content of home movies, even further contributing to their iconographic visual postures and tendencies toward stereotyping. And the need for special lighting equipment for indoor scenes prompted many photographers to choose outdoor events instead, requiring less guesswork to shoot with the proper exposure. If, therefore, we wish to explain the prevalence of birthday parties, smiling faces, and family barbecues common to so many home movie artifacts, we must not assume that these representations by themselves express the intentions of practitioners. Instead, we must also consider the material and economic constraints of the apparatus and substrate, which lend themselves to documenting happy times, special occasions, and ritual events— important representations to be sure, but not necessarily or wholly constitutive of home mode practice, particularly in the era of video.

Videotape's low cost, extended recording time, and capacity to be recycled substantially increase the potential range and volume of events and behaviors recorded during home mode production. Because long takes cost less and tapes need not be changed for hours at a time, practitioners may be less selective in advance of shooting. Synchronous sound recording, standard to all video equipment, introduces an extremely important technique missing in home movies: on-camera narration. Preventing the need for icons or stereotypes alone to express meaning, sound allows practitioners to comment on the action, tell stories, and interview subjects, increasing the likelihood of narrative content and interpretation within the artifact itself, rather than only as an external accompaniment at the site of exhibition. Video's sensitivity to low light levels also multiplies the number of indoor scenes recorded, which generally portray more quotidian activities lacking ritual intent. Indeed, since it may be left running for hours on a tripod, ultimately forgotten, the camcorder fits more easily into everyday life without intervening in routines, selecting content, or posing subjects. This seeming "transparency" of home video, its less self-conscious presence on the scene of domestic living, tends to relax some of the artificial conventions imposed by home movies, even when camcorders are appropriated to document special events. Therefore home videos may portray not only the birthday party itself but also baking the cake and wrapping the presents; not only smiling faces directed to say "cheese" but tears, boredom, or anger when subjects forget they're on camera; not only outdoor scenes of barbecues or trips to the beach but indoor scenes of cleaning, bathing, or just plain hanging around. While some home video practitioners may choose to erase or record over footage that fails to portray their families or friends in a positive light, just as many will cherish moments of embarrassment, distress, or defeat for their candid humor or truth. Video, that is, realizes a broader range of intentions than Chalfen's formal reading of home movies would indicate.

Like its production methods, home movie reception practices augment the ritual quality of its artifacts. Exhibition demands a special screen, projector, and, most importantly, darkness—all requirements that recontextualize reception as an unusual behavior outside of the everyday. Thus, watching home movies itself becomes a special event, and its anticipation is heightened all the more by the time lag between production and reception necessitated by lab processing. Although this separation between the moments of recording and viewing provides a space for distance and interpretation of the events on film, which may defamiliarize their ritual quality, because projection of home movies runs forward automatically without significant options for intervention, the flow of images in more or less real time may resuture this gap by charging the imagery with a larger-than-life significance associated with cinema in general. Finally, because expenses of time and money discourage duplication of home movies, practitioners tend to contain them within the home itself rather than distribute them to an extended circle of family and friends. Like most keepsakes specific to one particular household, home movies become imbued with sacred meaning.

Home video, on the other hand, closes the gap between production and reception. Its images may be viewed, in fact, while the event being recorded is still in progress, imbuing its artifacts with a self-conscious reflexivity foregrounding the theatricality inherent in the home mode itself. In playback mode, its long takes increase the spaces of ambiguity and therefore interpretation of events. Its familiar exhibition apparatus, a VCR and domestic monitor typical for domestic television reception, requires no special lighting or setup, making home video more continuous than home movies with the everyday activities it depicts. While this segue between life and image may naturalize home video artifacts, the VCR's controls may manipulate the flow of images: fast-forward, instant replay, fast motion, slow motion, reverse motion, freeze-frame, and frame-by-frame functions allow for intervention, analysis, and play, revealing the tape as an artificial construct defamiliarizing video's so-called reality effect. And finally, the ease of distributing videos outside of the domestic circle by simple, low-cost duplication within the home by means of two VCRs enlarges the artifact's audience, diffusing its sacred importance to one family. Indeed, as the popularity of *America's Funniest Home Videos* should indicate, audiences may extend from coast to coast.

These differences between home movie and home video production and reception practices, determined in part by differences between substrate and apparatus, must be taken into account to understand how video may not necessarily transform the home mode (which implies technological determinism) but may increase opportunities for representing a greater range of social intentions less likely to emerge on celluloid. Therefore, rather than assume that video in itself may "revolutionize" amateur practice because it changes conventional perceptions of domestic living (such as home movies), we more properly should conclude that the new medium is more likely to represent the fuller range of domestic ideologies already present in the culture, well before the arrival of home video or even amateur photography itself.

In short, home video reveals that families have always been more complex and contradictory than home movies have generally portrayed them. Let us note, for example, that much content absent from home movies appears in home videos. As Chalfen has defined the home mode, one of its primary characteristics is its set of "patterned eliminations" that select only a narrow, partial spectrum of everyday life to represent domesticity in its imaginary forms.[11] Thus, for example, he cites the absence of depictions in snapshots and home movies of preparing meals, repairing the house, watching TV, using the telephone, or working at one's place of employment as the effect of ideological choices to present an idealized image of home. Does the ubiquitous presence of these very activities in home videos suggest either that video has transformed these ideologies or that these ideologies have changed themselves over time? While the second possibility seems more likely, perhaps a better proposition altogether would suggest that video has liberated the constraints of the photographic apparatus pressuring home mode practitioners toward these eliminations, patterned as much by material and economic choices as by ideology. Some of Chalfen's other examples of patterned eliminations, such as scenes of intimate lovemaking, family quarrels, and death ceremonies, still typically absent from home video, may indeed be patterned by cultural prohibitions; on the other hand, amateur pornography, modeling therapy, and memorial tapes continue to break new ground in the field of video practice, suggesting that changes in home and family values may be working in tandem with changes in the technology itself, if not determined by them.

Of significant importance to any evaluation of the home mode will be the media analyst's theoretical interpretation of these "patterned eliminations." For Chalfen, a cultural anthropologist, they define home movies positively as signifiers of social intention and symbolic world making. For Zimmermann, a political modernist, they define home movies negatively as structuring absences of critical modes of thinking and avant-garde pursuits. Although ideology underlines both of their approaches, Chalfen never reaches Zimmermann's conclusion that the home mode reproduces false consciousness, that its eliminations of content are coerced by controlling corporate agencies. Instead he grants an autonomy to the home mode, which Zimmermann would rather hold accountable to more radical modes of practice.

▶──────────────────────────────────────

Home Mode Families: From Nuclear to Discursive

Both Chalfen and Zimmermann construct the "home" in their models of the home mode according to the sentimental precepts of domesticity limited mostly to the white middle-class nuclear families predominant immediately following World War II. This attribution does make sense methodologically, because during this period, snapshooting and home moviemaking rapidly increased in popularity as a leisure time pursuit within that demographic group. Nevertheless their models might locate their historical moment more specifically. Though Chalfen rarely specifies the particular makeup of the families he has researched, his conclusions about practice do imply that the nuclear model functions as a standard, although not by necessity. Zimmermann, on the other hand, explicitly defines home mode practices according to particular ideologies of familialism that predominated briefly, approximately from 1945 to 1965. Thus the nuclear family serves throughout her analysis as a synchronic model.

Zimmermann charges the home mode with reproducing the nuclear family by its instrumental diffusion of domestic ideologies that usurp and defuse radical amateur practices threatening the security of the status quo. The title of her study, *Reel Families,* implies false consciousness by suggesting that "real" family relations have been replaced with their beguiling representations in home movies, misrecognized as real, and thus acting as duplicitous agents of capitalism. While her Marxist approach is theoretically sophisticated and politically motivated, it is less illuminating when applied to families recorded in the era of home video.

What are the primary characteristics of the sentimental model of domesticity typical of home movie discourse? More or less, they align with the values and formations of the patriarchal bourgeois nuclear American family. Home is defined as a private place of leisure and consumption, marked by interpersonal cooperative values and gendered as feminine. Work is defined as a public space of labor and production, marked by impersonal instrumental values and gendered as masculine. By default, the vital functions of home are narrowed to bearing, rearing, and socializing children, each divided and regulated by specific sex role expectations: husband and father as autonomous professional and dominant authority, wife and mother as dependent homemaker and emotional caretaker. Because the structures of the competitive public sphere tend to segregate husbands from wives and fathers from children during working hours, leisure time pursuits cohere around two domestic ideologies: first, "togetherness," which promotes "the bourgeois nuclear family as the only social structure available for the expression of common, shared experiences that could shore one up against alienation and isolation," and second, "familialism," which transfers "the idea of the integrated family unit as a logical social structure onto other activities."[12]

Zimmermann argues that these twin ideologies indicate how social, cultural, and aesthetic formations became organized along patterns resembling those of this domestic model. As one of the most important forms of its own recreation, that is, the nuclear family adopted amateur film as a hobby in which all members could participate. To explain why potentially indeterminate amateur technologies might readily be appropriated for home movie practice, Zimmermann advances two hypotheses. First, the ideologies of togetherness and familialism exerted their effectivity from commercial discourses of advertising and training manuals. Second, these ideologies insidiously co-opted all amateur technologies and practices in their entirety: "This weaving together of amateur film as a hobby and familialism as an ideology of social interaction permanently displaced any other production possibilities for families to such a degree that any distinction between amateur film and home movies collapsed" (132).

Surely home movies, at their peak of popularity when the bourgeois nuclear family exerted hegemony in American society, must in many cases have reiterated the status quo, therefore justifying Zimmermann's critique to some degree. But her censure of home video on the same grounds is less compelling: "The home-video camera emerges as a silent relative at family gatherings, never interrupting, never gossiping, never interpreting as it records hugs, kisses, hamming, and idealized memories of a contrived family harmony" (150). This description seems to be a gross caricature of home video practice today, particularly because Zimmermann's "reel families" are giving way to "families we choose," whose political ties are more discursive than biological, diffuse than nuclear, diverse than monolithic, oligarchic than patriarchal. True, the sentimental model essential to Zimmermann's critique remains a powerful residual presence in the cultural imaginary of American families and may arguably continue to govern as the ruling image in advertising discourse. But it is only one model among many that now configure the emerging forms of family life, any or all of which shape the social imagination of contemporary amateurs who increasingly construct their idea of family by act of choice, rather than by kinship, filial, or spousal obligation. By archiving events as diverse as roommates on vacation, promotion parties at work, and second marriage ceremonies of recently divorced spouses, home video libraries document changing cultural conceptions of home itself.

In *Critical Theory of the Family,* Mark Poster cautions against studies that construct a history of the family as its inevitable evolution toward a configuration expressing the instrumental functions of capitalism.[13] Instead he advocates a diachronic description of the family's diverse and distinct structures operating within a complex social formation at any given time. Noting that the modern nuclear family emerges prior to industrial capitalism, Poster argues that family forms enjoy at least a partial autonomy from the state and the economy. He therefore rejects the critical tendency to reduce the specificity and agency of family experience to a functionalist expression or to uniform socialization and equilibrium, for such premises imply that present family forms cannot be changed in the future.

Following Poster's recommendation, we must update and revise ideologies of familialism to account for changing family forms. As many contemporary sociologists of the American family have observed, however valid it may have been in the 1960s to identify the suburban nuclear-family lifestyle as characteristic of the society, it was certainly not valid by the 1980s.[14] Indeed, rather than predominant in the twentieth century, the sentimental model has been a statistical aberration: "Far from being the last era of family normality from which current trends are a deviation, it is the family patterns of the 1950s that are deviant."[15] And as Steven Mintz and Susan Kellogg point out, "the high marriage and birthrates and relatively stable divorce rate of the 1950s were all sharply out of line with long-term demographic trends."[16] Since the beginning of the 1970s, the postwar processes that might have standardized the family life cycle slowed down and reversed, resulting in an increasing proportion of household configurations departing from the normative nuclear pattern. Although economic factors, such as the decline in overall prosperity and the rise of inflation, would exert their own effectivity by demanding, for example, that former housewives return to the workforce to maintain a standard of living, Arlene Skolnick contends that changes in family forms were also determined in part by changes in family ideology:

> By the early 1970s, traditional assumptions about family life had been undermined by both ideological challenge and behavioral change. The undoing of 1950s manners and morals begun by rebellious youth was spreading to mainstream America; the women's movement was reshaping public opinion and unsettling relations between men and women inside and outside the family. Meanwhile, the newly emerging gay liberation movement was challenging the norm of heterosexuality.[17]

Various social trends of the late twentieth century therefore severely compromise a transhistorical application of postwar "familialism" to home video practice: the general undermining of parental authority as the primary socializing agent by school and media institutions; a jump in divorce rates and serial marriages; rising rates of widowhood and the increasing economic power of women, both of which support the establishment of female-headed households; custody decisions that identify children as members of more than one household; the constitution of households by nonkin relations; the growing percentages of people living alone; and the heightened public presence of homosexual couples. These changes signify not only a breakdown of nuclear ideologies but new patterns of domestic living taking root in material culture.

Kath Weston refers to this new plurality of domestic patterns as "families we choose," whose members are adopted rather than ascribed: "Chosen families do not directly oppose genealogical modes of reckoning kinship. Instead, they undercut procreation's status as a master term imagined to provide the template for all possible kinship relations."[18] Although referring specifically to gay families, Weston's template may also describe any congregation whose members share a common set of cultural values transcending the ties of heterosexual procreation, and locate "home" in a variety of environmental contexts, including the workplace, neighborhood, and school. Thus, as well as nuclear families, we find vocational families, avocational families, educational families, and professional families, which continue to perform the functions of socialization, if not necessarily biological reproduction.

How prevalent is this discursive idea of family within the population at large? Skolnick notes that "in a study conducted for the Massachusetts Mutual Insurance Company that asked respondents to choose among several definitions of family, three-quarters rejected the more traditional versions, picking instead 'a group of people who love and care for one another.'"[19] These findings suggest that we might rethink the "home" in the home mode more in terms of "household" or "peer group," since families we choose comprise a variety of constituencies and moral economies untethered to the bourgeois nuclear model. Although we regularly refer primarily to blood relations when we speak of our "families" in vernacular English, when we construct our images of "home" in snapshots or video, we tend to focus on diverse relationships flowing from shared activities and sensibilities rather than only on biological kinship itself.

In summary, the home mode in the era of video should not, in fact cannot, be stereotyped and globally condemned as a practice reproducing uniform family ideology. While ideologies do continue to shape the home mode, they have changed considerably, evident within home video artifacts themselves. While it would be disingenuous to dismiss contemporary political and moral discourses committed to the nuclear family, we should consider much of its polemical rhetoric as a reaction to changes in family structure rather than as a reflection of dominant ideology. As videos of gay commitment ceremonies illustrate, in an era of families we choose, the home mode's "conservatism" may be redefined not necessarily as "reactionary" or "regressive," connoting its political functions serving patriarchal capitalism, but more generally as "reconstructive" or "restorative," connoting its ritual functions serving the need for meaningful community. In an increasingly mobile and diverse society, the home mode persists as a practice to trace and make sense of the life stories of transient individuals seeking interpersonal communion within emerging family forms. Therefore locating the sources and values of this desire for home and family, whether biological or discursive, will concern the remainder of this chapter's attempt to explain and understand the persistence of the home mode, despite changes from celluloid to video and reel families to families we choose.

▶

The Home Mode of Cultural Reproduction: From Ideology Thesis to Habitus

For at least fifty years, snapshot photography, home movies, and home video have persisted as the most popular and widespread forms of amateur media practice. I will attempt to account for this inertia by analyzing the home mode as a site of cultural reproduction, a concept that serves to articulate the dynamic process that makes sensible the utter contingency of, on the one hand, the stasis and determinacy of social structures and, on the other, the innovation and agency inherent in the practice of social action. Cultural reproduction allows us to contemplate the necessity and complementarity of continuity and change in social experience.[20]

I will limit the range of explanations to the two most prominent positions in home mode discourse, although such an approach is by necessity reductive. The first position argues in favor of dominant ideology: that hegemonic discourses appropriate media technologies to contain practice to family representation in their effort to reproduce the imaginary relations of capitalism. The second argues in favor of cultural praxis: that deeply structured habits, such as family representation, already embedded in material practices, appropriate amateur media technologies to express a coherent worldview consistent with the particular histories of a class of practitioners. Both positions leave room for ideological critique, but the former generally evaluates the home mode as reactionary, while the latter describes its practices as expressions of social intention marked by cultural distinctions of relative political value.

Incidentally, Richard Chalfen takes a third, centrist position. Describing the reproductive operations of snapshots and home movies as "Kodak culture,"[21] artifacts both about and of culture, he interprets the apparent redundancy of home mode imagery as a reiteration of culturally structured values and as a broad reaffirmation of themes of stability, conformity to norms, and maintenance of ethnocentric values (139). Inflected by ideas, values, and knowledge that are informally or unconsciously learned, shared, and consensually agreed upon in tacit ways, the home mode potentially transforms ongoing patterns of activity into other behavioral routines that are socially appropriate and culturally expected when cameras are in use (10–11). More specifically, Chalfen adds that home mode artifacts express and bequeath ideologies that positively value individual progress based on conspicuous success and acknowledgment that each step in life has been taken in the proper and conventional manner (99). Therefore he concludes that upon viewing snapshots and home movies, children internalize these memories of achievement and happiness with an unspoken expectation that this pattern should be repeated (140).

Although Chalfen affirms that the content and form of home mode artifacts vary in relation to historical time frame, family composition, religion, ethnicity, geographic location, class, race, and sexuality, his focus on North American practices clearly inflects his notions of Kodak culture with the values of capitalist democracy (97). He never concludes, however, that the home mode is completely determined by these ideologies, but rather expresses them as symbolic statements of consensus. We should therefore imagine that the home mode operating in a subculture structured by different socioeconomic values may express a different kind of consensus, even if operating within the wider sphere of industrial culture.

Such imagination is, however, rarely exercised in home mode discourse. Indeed, most critics of the home mode seem merely to have extrapolated from Chalfen's description of Kodak culture a set of prescriptive and proscriptive doctrines: that the home mode's unconscious motivations suggest false ideology; that its structured redundancy demonstrates the functionalist fit of its practices; that childhood internalization of its messages points to its power to socially reproduce the family in its own image; and that its emphasis on conspicuous consumption and individual progress testifies to its instrumental capitalist values.

■──

The Ideology Thesis

Arguments in favor of these premises that home mode practices and artifacts are both the determined products and determinate producers of capitalist values can be grouped under the "dominant ideology thesis," a form of interest theory that identifies ideological positions derived from the investments of various power blocs as antecedent causes of widespread cultural values and behaviors.[22] For adherents of this thesis, the Eastman Kodak corporation stands as one primary interest group disseminating the ideology that circumscribes amateur photography within the home mode. Such critics caricature Chalfen's "Kodak culture" by suggesting that the ideological patterns he describes have been pressed on practitioners by Kodak's monopoly on the design and marketing of amateur technologies.

Subscribers to the ideology thesis often privilege three types of discourse to make their points: advertising, manuals, and popular entertainment as a model for imitation.[23] From narrowly selected texts they hope will "prove" that these discourses have "trained" amateurs how to use their cameras,[24] these critics of the home mode extrapolate a set of codes and standards that they imply were invented by media corporations and enforced on the masses. Such conclusions implicitly deny that the desire to represent home and family may be embroiled in a history of material practices predating the designs of Kodak and its competitors, which instead appear to descend upon the masses from some epiphenomenal realm.

For example, Laurie Ouellette has employed the ideology thesis to explain the prevalence of video in the home mode. Observing that the vast majority of consumers have not used their camcorders in oppositional ways or sought out exhibition opportunities on cable access, she wonders: "Why is this so? If the problem is no longer access or affordability, it becomes a question of 'taste,' that is, exposure, conditioning, and ideology."[25] Rather than locate the question of taste within a material history of aesthetics, Ouellette presumes that "discourses such as advertising, news, and entertainment TV have consistently sought to shape the video revolution in ways that subvert real citizen input and uphold the authority of the existing media system."[26] Without outlining precisely what these discourses seek to gain by subverting "democratic" amateur practice, Ouellette presumes that their messages are somehow fundamentally at odds with some oppressed intention of the general population, who, if not for their coercion by dominant ideology, would take up their camcorders in the service of some yet unspecified "revolution." Ouellette plays down the possibility that manufacturers might be providing amateurs with equipment designed to fulfill their very real, deeply structured desires for family representation, which instead she implies must be the effect of "conditioning." Thus her explanation for the predominance of home video mirrors Zimmermann's about home movies: "The marketing emphasis on home mode representation over all other possibilities has limited the democratic potential of widespread camera ownership."[27]

Ouellette's argument raises many unanswered questions. How, exactly, is home mode practice a compromise of democracy? Why demand that amateurs seek cable access, an imperative that replicates journalistic and entertainment discourses dictating that only the most worthy video finds exhibition on television? What ethnographic data supports her declarations about the majority of amateurs, whose millions of tapes remain cloistered from her observation? If, as she suggests, activists and progressive media critics have managed to liberate themselves from victimization by advertising messages, what evidence or theory proves that the remainder of the population is therefore more prone to discursive conditioning?

Ouellette adopts the ideology thesis as a progressive media doctrine without interrogating its following fallacies: (1) an assumption of a necessary coincidence between the effects of capitalist processes (advertising) and the ideological needs of a dominant field of practice (the home mode), (2) an implicit appeal to false or repressed consciousness that denigrates home mode artifacts as second-order representations, and (3) a faith in spontaneous subjectivism, by which amateurs, once limited from dominant ideology and "domesticated" technology, will automatically dispense with the home mode to volunteer as grassroots video activists. In short, according to the ideology thesis, if it weren't for Kodak commercials and their ilk, we'd all join Michael Moore's "TV Nation."

Perhaps the most fundamental problem with the ideology thesis is its blind spot presuming a one-to-one correspondence between discourse and practice. Ouellette offers no ethnographic evidence from empirical home video artifacts to prove that commercial discourses have been the leading cause for the predominance of home mode practices. Zimmermann at least does offer a qualification of her study's underlying premise: "This analysis of amateur-film discourse raises the inevitable question of just how much the discourse of advice columns and camera manufacturers overdetermined the actual practice of amateur filmmakers."[28] Yet without offering a definitive theoretical answer or a significant sample of amateur artifacts, she proceeds to speculate as though discourse and practice were identical.

Perhaps a true test of the ideology thesis would be to imagine the effects of an advertising discourse advocating antifamilial practices: if Kodak were suddenly to air commercials portraying amateurs taping pro-choice rallies, child abuse, and adulterous affairs, would the home mode give way to these practices? Richard Chalfen himself questioned the efficacy of external professional or commercial recommendations for home mode practice, observing that "patterns of on-camera behavior contradicted the behind-camera objectives recommended by the manuals."[29] Indeed, for Chalfen, the home mode conforms to folkways opposed to institutionalized norms. Pierre Bourdieu, in his study of photography, concurs: "Photographic practice is considered accessible to everyone, from both the technical and the economic viewpoints, and those involved in it do not feel they are being measured against an explicit and codified system defining legitimate practice in terms of its objects, its occasions and its modalities."[30]

At bottom line, the ideology thesis betrays the fallacious logic of expressive causality outlined in chapter 1: home mode artifacts express and serve the ideologies of the dominant social order.[31] Rejecting this kind of facile equation, Raymond Williams writes:

> Theoretically the problem is that the "social order"—here a formal term for social and historical process—has to be given an initially structured form, and the most available form is "ideology" or "world-view," which is already evidently but abstractly structured. This procedure is repeated in the cultural analysis itself, for the homological analysis is now not of "content" but of "form," and the cultural process is not its active practices but its formal products or objects.[32]

For subscribers to the ideology thesis, who derive models of the dominant social order from selective marketing discourses and the nuclear family, home video can therefore only be an expression of advertising and familialism.

If it ever did in the past, this equation fails to balance in the present. As Christine Tamblyn has properly observed, "the major corporations that manufacture camcorders no longer aim their marketing campaigns solely at the bourgeois heterosexual family."[33] Even a cursory glance at trade journals and video handbooks supported by corporate advertising will bear her out. *Video Review,* for just one example, has hosted an annual "Shoot-off" competition, awarding prizes to nonprofessional videotapes in categories ranging from documentary and environmental issues to special effects and animation. "Family fun" is only one category among many. Some of the winning tapes included a lampoon of PBS programming, a moody tone poem, and a music video about ambiguous sexual identity. Far from "training" amateurs to replicate conventional home movies, the editors have declared, "as the camcorder evolves from exotic toy to household staple, we'll be seeing more and more work along these lines."[34]

In another example, *Video Maker* magazine's regular column "Carry a Camcorder" once encouraged amateurs to shoot home video diaries. The writer's only advice about following standardized codes was to avoid them altogether: "The first thing to remember is that video of this sort needn't follow any preset rules."[35] And the writer not only discouraged narrative coherence but advocated recording family arguments and heated discussions.

As these examples illustrate, contemporary home video discourses have encouraged amateurs to make do with consumer technologies from a variety of ideological perspectives. Even those who continue to document home within the conventions of familialism can never guarantee that their artifacts will express or reproduce its ideologies. Take, for instance, Michelle Citron's *Daughter Rite* (1978), a film that interrogates the oppressively sexist values encoded in her father's home movies. By denaturalizing their images through techniques of optical step-frame printing, which opens up the potential for close analysis and subsequent criticism, Citron defuses the movies' ideological effectivity through an operation of deconstructive decoding. Her film visually materializes the operation by which individual spectators may actively negotiate even the most reactionary values of the home mode rather than passively absorb them. While parents may intend their home movies and videos to signify togetherness and happiness, their children may contest or resist these messages by interpreting them from the different perspectives of their own personal experiences. Therefore the meaning of any home mode artifact may change over time, preventing a uniform or functionalist fit between family members, home mode practice, and the social order.

In summary, as a top-down approach, the ideology thesis succumbs to the fallacies of mechanistic determinism, interest theory, and expressive causality. It conflates discourse and practice. It condescends to the masses as dupes of false consciousness. It aligns all ideologies of family with those of the dominant social order. It denigrates home video as a deformity or betrayal of amateur practice. Its deductions, generally unsupported by ethnographic evidence, are difficult to verify. Let us turn, then, to an alternative thesis with greater exegetic power.

Habitus and Field

A bottom-up approach that argues for the effectivity of ideologies born of and embedded in material practices, Bourdieu's theory of practice locates the origins and persistence of the home mode in the historical experiences and cultural environments of its practitioners rather than in epiphenomenal discourses.[36] However reactionary or innovative the message that these discourses communicate, their effects will be incorporated, modified, or rejected by habits already made. Thus the ideologies of practice advocated by advertising and training manuals will be adopted only if they find social effectivity within the life worlds of the subjects they might hope to interpellate. Denying that ideologies abstractly or mechanistically create wants or needs, Bourdieu emphasizes that the ideological contents of consciousness, whether of the dominant or the subordinate, are less relevant in explaining behavior than the availability of resources that orient, constrain, or sustain media practitioners.

Rejecting mechanistic objectivism and spontaneous subjectivism, twin fallacies of the ideology thesis, Bourdieu introduces two key dialectical terms for explaining the forms and content of any cultural practice. "Habitus" reintroduces notions of human agency denied by structuralism while avoiding the free-floating pluralism often celebrated by cultural studies. "Field" grounds agency within structured social relations without succumbing to determinist objectivism, which overlooks the historically specific cultural conditions of distinct practices. Together, habitus and field describe the relationship between observed regularities of social action (structure) and the experiential reality of purposeful human actors (intention). For Bourdieu, any field of practice is structured by social relations, cultural functions, and environmental context, each of which requires a historical explanation to understand under what conditions that field may be transformed or reproduced.

In particular, Bourdieu's concept of habitus explains why the home mode has prevailed as the predominant amateur media practice since the invention of photography. Defined as the inculcation of a system of transposable dispositions that integrates an inherited set of meanings and codes (cultural capital) specific to a family, group, institution, or class with the material resources (economic capital) available to its members, a habitus functions as an analogical matrix of schematic perceptions and behaviors that inflect strategies for the future and generate a subjective judgment of practices as an objective classification. This structuration of practices and representations can be regulated without being the product of obedience to discursive rules and can be adapted objectively to a set of goals without presupposing a conscious aiming at ends or an express mastery of the operations necessary to attain them. This lack of self-consciousness does not indicate false consciousness as much as the determinations of past conditions that have shaped the field in question.

Thus, for example, Bourdieu accounts for the home mode of photography not as the invention of camera manufacturers, as does the dominant-ideology thesis, but by prevailing cultural inertia: "The photographic image . . . came to fulfill functions that existed before its appearance, namely the solemnization and immortalization of an important area of collective life."[37] Thus habitus, like medium specificity itself, is a mediating discourse that generates and validates practice as proper to the goals defining it: "The habitus is the universalizing mediation which causes an individual agent's practices, without either explicit reason or signifying intent, to be none the less 'sensible' and 'reasonable.'"[38]

Although Bourdieu more generally theorizes the habitus of social class, one could apply his concept to a different notion of "class" at the institutional level.[39] For example, the habitus specific to a contemporary art museum, in part constituted by the material need to exhibit installations and to seek funding from granting foundations, tends to regulate video practice as "video art," an intentional disposition adapted to the goals necessitated by the structure of the museum: "The homogeneity of habitus is what—within the limits of the group of agents possessing the schemes (of production and interpretation) implied in their production—causes practices and works to be immediately intelligible and foreseeable, and hence taken for granted."[40] In sum, the habitus exerts a probabilistic logic derived from a common set of goals, expectations, and material conditions to regulate the practice of a set of individuals in common response to their shared social environment and cultural history.

Within the field of home mode practice, the family may be theorized as an extremely localized class sharing its own habitus shaped by a logic internalized in early childhood through a given set of conditions mediated through the dispositions and practices of adults. Home mode artifacts function, therefore, as material expressions of schemes of perception, thought, and appreciation common to the family as a group. Its norms that organize and evaluate what is worthy and not worthy of being represented are entwined in a local system of values that cannot monolithically be reduced to the dominant ideologies of a more global social order, in particular during the contemporary era of families we choose, each based on a constantly shifting set of biological, social, and discursive relations structured by a habitus seeking the common denominators shared by all members. As every practice is conditioned by a habitus biased by predispositions and hierarchical value judgments, none will be freer of ideology than another, more authentic than another, more real than another, but in competition within a shared field.

Bourdieu's theory of practice allows us to understand the patterned eliminations of the home mode as neither only what its artifacts proclaim nor only what they disguise, as neither a transparent nor a deformed practice. It reestablishes the relative autonomy of the home mode within the social order, not as a self-identical practice immune from its historical relations, but because its cultural functions, depending on the communities that they serve, may or may not express dominant ideology. Although each family may appropriate amateur technologies for idiosyncratic purposes unique to the aims of its members, the history of home mode conventions may also structure their practices according to functions more broadly shared by large numbers of families crossing various differences of class, environment, and constituency.

My attempt to restore the home mode to its own habitus requires that I point out its positive cultural functions rather than bemoan those that are absent. Only then may we understand its persistent autonomy and popular attraction. Before doing so, I need to distinguish the term "cultural function," a description of a structured set of dispositions and intentions, from the "functionalism" and "strain theory" often proposed as explanations for the home mode's inertia. A functionalist approach would locate the home mode within the logic of expressive causality as an institutionalized practice fitting seamlessly into the culture at large. Like a microcosm of the social totality, the family in a functionalist paradigm must reproduce its operations from one generation to the next. The home mode, as a mediating agent of that reproduction, would emphasize family stability, consensus, and continuity to preserve the status quo. Although this is an attractive argument, primarily because it makes for a clean theory of social formations unmuddied by contradiction, even the most cursory empirical analysis of variegated family lifestyles will belie its persuasiveness. Neither can a functionalist paradigm of the home mode properly account for individuals such as Sadie Benning, Michelle Citron, or George Kuchar, whose avant-garde domestic representations of home and family would be evaluated as dys-functional rather than revisionary.

Strain theory, a variant of functionalism, accommodates the contra-dictions within a social totality by defining ideology as a mechanism for restoring equilibrium unbalanced by the conflicts of modern life.[41] Within this paradigm, collective structural alterities, such as the demands of work opposing the desire for leisure, would trickle down to the individual as a personal struggle between the norms and goals of the economy and those of the family. According to strain theory, therefore, the home mode would function as an improvised defense against frictional strain and a symbolic agent for releasing tension. Providing a relaxing activity that celebrates do-mestic autonomy as it records acts of conspicuous consumption, it would satisfy the demands of both leisure and work. Because subjects of capitalist ideologies all suffer similar strains, so too would their home movies and videos share similar content, explaining their redundancy from family to family. Although strain theory does offer one explanation for the home mode's emphasis on community—that it provides the solace of solidarity in a rationalized, fragmented, individualistic, competitive society—it mistakes home mode functions as an effect of the contemporary order rather than understanding them as rooted in kinship traditions predating capitalism.

Functionalism and strain theory have been attractive to critics who wish to link the home mode expressly to capitalist ideology. In particular, because they assume that its artifacts passively reflect bourgeois values, such critics deny the home mode's status as an active, creative mode of pro-duction and instead degrade its practices to mere acts of consumption.[42] This position often depends on a combination of the following three argu-ments. One, because its automatic technologies passively record and trans-mit found reality rather than transform it, the home mode supposedly consumes rather than creates the world. Two, as a leisure pursuit strictly opposed to work, the home mode typically documents only consumer be-havior and commodities (gift giving, vacations, wedding receptions). Three, because home mode practices require an initial investment in goods and services (e.g., camera, film processing, prints), their operations depend more on consumer choices than on a deliberate aesthetic.

While these arguments don't ring entirely false, neither do they apply only to home mode practice, whose material specificities merely foreground the primary paradox of contemporary popular culture: creation through mutual acts of production and consumption.[43] To condemn the home mode for its dependency on acts of consumption would therefore be to condemn virtually all creative practices within a consumer culture. Such arguments cannot be sustained since the beginning of the twentieth century, after which the fields of amateur and commercial practice have been regulated by their ever changing mutual relations. In particular, beginning with photography in the late nineteenth century, the home mode has always been defined in its relation to commerce, from which it was born. To wish it as a purely autonomous folk subculture would be woefully naive. The fundamental error is to oppose production and consumption as wholly antithetical activities. This polarity can only construct arbitrary hierarchies that ignore the acts of production and consumption within any practice.

Home mode artifacts transcend their initial origins as partial commodities once they have been appropriated by, and incorporated into, the household that imbues them with personal meaning:

> Households are conceived as part of a transactional system of economic and social relations within the formal or more objective economy and society of the public sphere. Within this framework households are seen as being actively engaged with the products and meanings of this formal, commodity- and individual-based economy. This engagement involves the appropriation of these commodities into domestic culture—they are domesticated—and through that appropriation they are incorporated and redefined in different terms, in accordance with the household's own values and interests.[44]

Here Roger Silverstone, Eric Hirsch, and David Morley describe the process of conversion, whereby consumer media technologies transform their potentially alienating commercial identities into expressions of a household's sense of itself and place in the world. As literal audiovisual displays of the household's self-image, home mode artifacts circulate as symbolic statements about membership, identity, and lifestyle within a private sphere of gift, rather than exchange, relations. Thus although produced in part by acts of consumption, snapshots, home movies, and home videos cannot be identified simply as commodities once their various means of production move from the formal economy of the market to the moral economy of the household.

Like the life histories that they record, home mode artifacts follow their own intimate trajectories, of which consumer necessity constitutes only one phase, neither pregiven nor inherent to their biographies over several generations.[45] Indeed, their accumulated symbolic value may completely offset their commodity value. Therefore, as they mark the passage between two phases of cultural identity—between exchange and gift value, between market and household economies, between objective material and subjective desire—snapshots, home movies, and home video provide their practitioners with valuable documents negotiating these tensions.

A Functional Taxonomy of the Home Mode

The purpose of this taxonomy is to evaluate the home mode in its positive relation to the social order, rather than its negative reflection, and thereby instate home video as a distinctive amateur practice with significant and valuable cultural functions. At the most fundamental level, *the home mode provides an authentic, active mode of media production for representing everyday life.* Because home mode practitioners are personally involved behind and in front of the camera, and deeply invested during exhibition, they exercise a vital role in all aspects of production and reception, perhaps more so than in any other media practice available to them. Simply put, the home mode can be a creative art, not merely a consumer reflex. Rather than passively aim and shoot, practitioners typically control the content of their artifacts, but according to social orientations more than aesthetic manipulations. To confuse the two may lead to specious reasoning.

For example, Richard Chalfen has noted how many scholars have carelessly attached the label "naive" to home mode practitioners, claiming, for instance, that their lack of technique betrays little concern for composition.[46] For the stylized compositions typical of art photography, perhaps. But what about different notions of composition based on subject rather than form? Think of the typical customs for taking a snapshot: making sure that everyone fits in the frame, that members turn forward to reveal their faces, that the taller stand behind the shorter, that fingers or other stray objects do not block the lens. Here composition serves to represent the subjects' identities in their wholeness, and while its aesthetic might seem transparent, perhaps clarity of subject matter is precisely one governing intention of the home mode's strategies for shooting.

To reduce the home mode's aesthetic to naïveté implies ignorance and lack of consciousness, which together may support claims for its practitioners' ideological false consciousness as well. But this slippery-slope deduction can easily be refuted by the examples of artists such as Ross McElwee, who self-consciously moves in and out of the documentary and home modes, each aesthetic deliberately chosen to signify an idea or dramatize a conflict well suited to that particular mode. Would we characterize the home mode moments in his work as "naive"? Probably not. But to reserve conscious intention only to "artists" at the expense of "amateurs" betrays an elitist snobbery.

Rather than transparent, naive, lacking in consciousness, or worse, reflecting false consciousness, the home mode's aesthetic serves a set of ritual intentions negotiating personal conviction and its submersion into communal consensus. Thus a second function of home mode artifacts is *to construct a liminal space in which practitioners may explore and negotiate the competing demands of their public, communal, and private, personal identities.* It is hardly coincidental that home mode practice gravitates toward documenting rites of passage themselves, pivotal events pointing in two directions at once. From birth to death, individuals seek to record their unique stories, but staged within a broader community narrative. The liminal form of home mode artifacts—material and immaterial, objective and subjective, conventional and idiosyncratic—is well suited to its liminal content, for it provides participants with flexible tools for enacting and re-enacting ceremonial behaviors, for moving between private experience and its public display, and for understanding unique identity within the culture at large. Unlike avant-garde practices that celebrate a purely idiosyncratic vision, the home mode documents the desire to lose ourselves in the group, and to discover what that means to our identities as individuals. It stages the hegemonic pull between agency and conformity in artifacts whose ambivalence will never be resolved by disinterested formal readings, but only by its invested participants who share a common history and life world.

This desire to merge personal and public identities is matched by a desire to merge past, present, and future generations. Thus a third function of the home mode is *to provide a material articulation of generational continuity over time.* In our contemporary era of families we choose, for whom traditions and conventions may be in continual flux, the home mode offers an important tool for tracing common roots no longer nourished only by blood. The act of videotaping, for example, becomes one such tradition, signifying an intention to link unique synchronic special events in a conventional diachronic series: "Traditions are represented as the means by which our own lives are connected with the past. Tradition is the enactment and dramatisation of continuity; it is the thread which binds our separate lives to the broad canvass of history."[47]

Traditions depend on shared values persisting through time. These values may encompass any moral and ethical principles participants wish to conserve. Whether right or left of the political spectrum, videos of KKK meetings or gay pride parades intending to celebrate and reconfirm the transmission of group values from generation to generation fall squarely into the home mode. Certainly, individuals may object to different sets of communicated values, but we should not let local evaluations hamper an analysis of a field of practice more generally concerned with merging individual and group identity in shared patterns of meaning and behavior persisting beyond the single mortal life span:

By paying attention to the intentions of others, to the goals of larger systems, one "buys into" a self that transcends the fragile differentiated individual. Thus as one grows older, the signs of integration tend to become more highly valued and consequently attract more attention: Objects that stand for memories, relationships, family, and values become more prominent.[48]

Home mode artifacts function not only as such objects themselves; they record and organize other objects in a mnemonic system that may literally be bequeathed to, and inherited by, succeeding generations.

Locating our position along this temporal continuum provides the home mode's fourth cultural function: *it constructs an image of home as a cognitive and affective foundation situating our place in the world.* Irreducible only to family, to household, to site of residence, or to place or origin, "home" is an ideal envisioned as the synthesis of three experiential domains: a personal, private space for memory and solitude; a social, public place for family or group interaction; and a physical environment designed for comfort and security.[49] A product of practical and emotional commitment to a given space, home is a phenomenological reality produced as the result of productive and reproductive work by its members to forge identity and maintain security.[50]

While usually thought of as geographic, home may be photographic as well, unconfined to a specific place, but transportable within the space of imagination. Thus the productive work of home construction may also include home mode artifacts traveling from country to country, city to city, or house to house as a set of gestalt images providing a cognitive point of orientation in a mobile society. For every secretary and executive who makes himself or herself "at home" in the office by decorating his or her desk with snapshots and framed portraits, the home mode constructs schematic reminders of their dwelling places of being.

If it constructs our sense of place, the home mode also chronicles our sense of history. Thus a fifth important function links the home mode to a long tradition of folklore and autobiography: *it provides a narrative format for communicating family legends and personal stories.* By focusing on the process of the family's ongoing creation and extinction—the addition of new members by birth, the loss of existing members by death, the establishment of relationships through marriage, work, or friendship—the home mode acknowledges the ephemera of individual life cycles while preserving them for posterity within a larger family biography. Thus in every home video, for instance, we will find a tension between the autobiographical and the historical, between a conscious desire to tell our own stories in the present and an unconscious concession that others will tell them about us in the future.

In a society in which we are increasingly confronted with images of others on billboards, magazine covers, and TV broadcasts, the home mode allows us to produce and circulate our own images, to measure them against these other images, and to negotiate their place in a mediated culture. As an audiovisual narrative, then, home video provides two autobiographical functions: to represent the events of one's own life, and to observe one's own image in action. While offering these opportunities for self-determination, at the same time, the home mode tends to conventionalize unique autobiographies as an allegory of "the life cycle," as an iteration of common human experience. Again we see another motive for its focus on rites of passage, which provide a fundamental narrative structure linking the particular and universal:

> Rites of passage are a category of rituals that mark the passages of an individual through the life cycle, from one stage to another over time, from one role or social position to another, integrating the human and cultural experiences with biological destiny: birth, reproduction, and death. These ceremonies make the basic distinctions observed in all groups: between young and old, male and female, living and dead.[51]

As a system of constructs that categorize and connect events and human identities through contrastive judgments of similarity and difference, the home mode imposes narrative order on an uncertain future by mapping out an itinerary model of behavior. If home video lets us see ourselves as ourselves, by recording the rites of passage made by other family members in whose steps we are encouraged to follow, it also lets us see ourselves in others.

This taxonomy of cultural functions and ritual intentions should not be construed as either exhaustive of the home mode or universally constitutive of all local practices. Indeed, for aspiring filmmakers who dream of making it big in Hollywood, an industrial model rather than the home mode will guide their amateur video practice. But for the heuristic purpose of establishing a foundation for the arguments of the chapters that follow, a middle-level definition of video in the home mode allows us to distinguish its mode of amateur practice from its sibling modes so that, although sharing family resemblances, each represents the expression of cultural functions for which the others should not be held accountable. In this way, we uphold the home mode not at the expense of other modes but as worthy of sharing in the field of amateur practice. This epistemological approach will allow us, for instance, to better identify the positive connotations of home modal moments that appear in even the most avant-garde films and videotapes without condemning them as frivolous or reactionary lapses in taste.

Neither true nor false, real nor unreal, the home mode communicates a meaningful view of *a* world rather than *the* world. To suggest otherwise is to deform the intentions of its practices and to condescend to the intelligence of its practitioners. As Sol Worth has advocated, "the study of culture is not accomplished by pitting symbolic worlds against one another."[52] To do so creates distinctions in the interest not of knowledge but of power. We turn now to a more in-depth interrogation of the costs and benefits of individuals and groups invested in marking such distinctions in the field of amateur video practice.

[3] *Modes of Distinction:*
The Home Mode, the Avant-Garde,
and Event Videography

> *Usually the amateur is defined as an immature state of the*
> *artist: someone who cannot—or will not—achieve the mastery*
> *of a profession. But in the field of photographic practice, it is*
> *the amateur, on the contrary, who is the assumption of the*
> *professional: for it is he who stands closer to the noeme of*
> *Photography.*
> :: Roland Barthes, *Camera Lucida: Reflections on Photography*

> *Art and cultural consumption are predisposed, consciously*
> *and deliberately or not, to fulfill a social function of legitimat-*
> *ing social differences.*
> :: Pierre Bourdieu, *Distinction: A Social Critique of the Judgements*
> *of Taste*

By the end of the twentieth century, American families had paid special-event videographers more than a billion dollars to record their rites of passage: births, bar mitzvahs, graduations, weddings, retirement parties—even deaths. Economic forecasts predict revenues will continue to grow indefinitely, suggesting that a novice practice formerly smiled upon as a passing fad now commands serious attention as an important business trend.[1] A local cottage trade during the 1980s, event videography has expanded into a national network of professional associations led by pioneering figures, whose high expectations for increasing financial profit and cultural esteem extend well into the new century.[2]

This network of entrepreneurs has emerged as an association of pseudoprofessionals whose practice simulates the home mode while diffusing it with instrumental values. Although the distinction between amateur home video and professional event video defines this new industry, fundamentally both categories of video artifacts share only one primary difference: home video of special events is produced for a community by one of its participants for free, while event video is produced for a community by an outsider for a fee. In most cases, both home and event videographers share the same technologies, the same domestic ideologies, and the same middle-class status. Although event videographers frequently attempt to distinguish their product by selling their skill at polished technique, the vast majority have no more formal training than the amateurs they seek as clients. Indeed, we may be hard pressed to tell the difference between an amateur and professional wedding video based only on a formal reading of the tapes themselves. And in fact, until the recent popularity of reality-based TV programs showcasing home video pressured the FCC to alter its broadcast standard laws, event videos, although supposedly professional, were labeled "amateur" and disqualified from broadcast.

Event video, therefore, is a puzzling phenomenon: industrial in value, amateur in status, professional in aspiration. Self-consciously positioning themselves somewhere between the fields of home video and industrial television practice, event videographers appeal to the discourse of professionalism to negotiate their amateur standing in the market at large. It is this discourse that marks a social distinction between home and event video, that carves out a commercial niche within the field of amateur video practice, that rationalizes the event videographer's payment for services rendered, for the event videographer's means of production—consumer equipment, low-end capital resources, and noncorporate credentials—fall squarely outside the means of dominant industrial practice and inside the field of the amateur. Unable to appeal to the same level of professionalism as network producers, event videographers must willfully create themselves as professionals by a different set of criteria in order to set themselves off from home videographers and to justify charging for their services. This odd status reveals two important insights about professionalism as an analytical category: one, that it should not be equated monolithically with dominant industrial practice, since both may diverge at the material and economic base of production; and two, that professionalism should therefore be conceived more as a discursively constructed, often arbitrary cultural distinction territorializing domains within a field of practice to demarcate a class of "legitimate" practitioners whose privileges include economic exploitation and increased social prestige.

As a hybrid of amateur, professional, and industrial practices, event video functions as an interesting case study to demonstrate the importance of defining categories of practice precisely. How exactly shall we classify event video? As professional? If so, how do we account for its material and economic differences from corporate television? As amateur? Then how do we account for its commercial exchange value? As a genre? How would we distinguish its textual signifiers from home video? As a mode? Then are its cultural functions distinct, or merely simulations of the home and industrial modes?

In this chapter, I will make an effort to sort out and define the categories of amateur and professional, not only to interrogate the complex operations of event video but equally to understand how the denigration of home video may itself arise from a misunderstanding and a misuse of these categories. The misunderstanding arises primarily from theorizing the mode of amateur practice as a genre rather than as an economic relation. That is, while the category of "amateur" should properly be conceived of as an umbrella term describing any nonindustrial media practice independent of the market or free of commercial exchange value, too often it is mistaken as a set of textual signifiers, techniques, and sociopolitical ideologies negating those of the industrial system. This approach tends both to essentialize and to politicize amateurism by equating its practices and texts with the avant-garde. Consequently, critical approaches to amateur practice will be inflected by a priori concepts—for if the avant-garde becomes the defining essence of amateurism by dint of its resistance to all things industrial, any amateur practice that overlaps with industrial techniques, properties, goals, or ideologies may be deemed corrupted or deformed. Here we see how the misunderstanding of amateurism leads to its misuse: a descriptive economic category slips into prescriptive aesthetic and ideological judgment and holds all practices within the amateur field accountable to the avant-garde. The home mode, therefore, as a practice that generally fails to interrogate dominant ideology in any explicit manner, may be condemned according to this logic as a betrayal of amateur strategies of resistance.

A primary aim of this chapter will therefore be to refine the category of amateurism so that we may restore its modes of practice to a state of relative autonomy and thus dispense with an either/or politics that must uniformly condemn one mode at the expense of another. In particular, as the relation between the amateur and the industrial fluctuates continuously over time, any critical approach based on identifying essentially amateur technologies, techniques, and aesthetics will be subject to ahistorical taxonomies. For example, although Hi-8 video may initially have been designed for and used primarily by amateurs, it was increasingly adopted by professional journalists; although jump cuts may have appeared experimental in the 1960s, they have become a staple of advertising and music videos; and although formal self-reflexivity may at one time have foregrounded a self-conscious awareness of the means of production, it has become popularized as a form of distraction in reality-based programming such as *The Real World.*

To rethink the amateur, the industrial, and the professional, I will proceed by redefining the amateur as an economic relation and delineating its modes of practice according to functional vectors that distinguish their separate aesthetics and cultural intentions. I will then trace the source of the conflation of the amateur with the avant-garde to important figures, including Maya Deren, Jonas Mekas, and Stan Brakhage, whose various comparisons of the amateur to both the avant-garde and the home modes have led to subsequent critical arguments holding the latter accountable to the former as a corrupted amateur practice. After deconstructing some of the myths of radical practice that have been marshaled to condemn the home mode, I will conclude with a detailed analysis of event video, whose hybrid status illustrates the fallacies of techno-aesthetic approaches to amateur and professional practice, since its artifacts simulate the home mode within the realm of commodity relations.

Underlying this analysis is Bourdieu's theory of practice, whose concept of "distinction" theorizes that oppositions between the amateur and the professional and between the avant-garde and home modes reveal elitist power plays in the service of social privilege and class hegemony. I will apply his notion of distinction to the concept of medium specificity in three ways: (1) in a heuristic sense of constructing taxonomies between modes of amateur practice for greater analytical clarity, (2) in a positive sense that each amateur mode should be recognized as a self-governing practice unaccountable to the functions of its adjacent modes, and (3) in a negative sense of elitist discrimination marking differences of taste and skill in attempts to territorialize and prescribe practice.

▶
_____

Refining Amateurism: Economic Relations and Functional Modalities

Fundamentally an economic relation, amateurism accommodates any non-industrial practice pursued for reasons other than market exchange. It is the presence of commodification that properly defines the industrial, and inversely the absence of commodification that defines the amateur—rather than the historical contingencies of their respective technologies, aesthetics, and ideologies. Thus, as a noncommodified practice, the home mode properly falls into the field of the amateur, and its cultural functions should be studied as relatively autonomous. Granting relative autonomy to the home mode complicates, revises, or outright rejects arguments that amateur practice should "rightfully" be avant-garde in nature. For example, "properly" amateur media artifacts are often described by media progressives as autotelic: anti-utilitarian, removed from exchange relations, and opposed to standards. Although home mode artifacts fail to meet these criteria—utilitarian in their use for ritual functions, embedded in relations of gift exchange, and measured by standards of realism—home mode practice is nevertheless exercised in and for itself, autotelic according to a different set of pleasures.

To restore amateurism as an economic relation and its various fields of practice as relatively autonomous, we must distinguish two analytical categories: "genre" and "mode." Whereas genre taxonomies generally identify differences among artifacts according to repeated patterns of internal textual signifiers (icons, thematic conflicts, narrative cues), modal taxonomies identify differences among practices according to repeated patterns of external cultural functions. Let us take documentary as one example to illustrate the difference. Although documentary is commonly classified as a genre in video rental stores, film guides, and college course syllabi, what common icons, themes, and cues do all documentaries share that would establish them as a coherent genre? Perhaps if we focus on one period or producer of documentary, such as John Grierson's work in the thirties, we might be able to list a catalog of shared elements: black-and-white footage, advocacy of social reform, sober narrative voice-over. Already, however, we can predict that by the time we arrive in the latter half of the century at the work of Chris Marker, this catalog will be revealed as ahistorical and obsolete: a residual definition failing to apprehend emerging documentary forms and content.

If instead of genre, however, we conceive of documentary as a mode (or modes) of practice, we may discover common underlying cultural functions that most, if not all, documentary artifacts in some form or another may fulfill, independent of the medium in which they are produced, the aesthetic of their techniques, and the substantive content of their subject matter. That is, rather than attend to the internal aspects of a documentary's signifiers, we turn to the external intentions of its practitioners to construct a taxonomy of documentary practice. Michael Renov offers an excellent illustration of the benefits of this type of functional modality, outlining four of documentary's primary cultural functions as (1) to record, reveal, and preserve; (2) to persuade and promote; (3) to analyze or interrogate; and (4) to express.[3] Stated in an infinitive form, these functions serve at once as nouns and verbs, signifiers of documentary as both artifact and intention, as a product of technology, aesthetics, and agency. Renov's functional modalities serve to widen the field of documentary practice to include a greater range of texts no longer limited by the rigid either/or binaries between nonfiction and fiction that would uphold, say, a Fred Wiseman at the expense of an Errol Morris.

In the same manner, by establishing a functional modality of amateur practices, we may widen the rigid either/or binaries that frequently uphold the avant-garde at the expense of the home mode. We must keep in mind, of course, that taxonomies based on functional modalities of practice do not produce uniform taxonomies of media texts themselves: several modes may operate within a single text. Yet unlike a hybrid genre, in which elements may mix, we would never speak of hybrid modes; instead, modalities tend to alternate within a text, each serving the cultural function for the moment it is intended. For example, returning briefly to documentary, Vivian Sobchack notes that the killing of a real rabbit in Jean Renoir's *The Rules of the Game* (1939) ruptures the film's symbolic representations with an extracinematic indexical referent; in other words, a documentary modal moment representing an actual death is embedded within a fictional narrative, which gains power from that moment's "ferocious reality."[4]

As Sobchack's example illustrates, the point of functional taxonomy is not to construct arguments about purity (something to the effect that Renoir compromises fiction with an excess of realism) but to locate and understand the diversity of intentions crisscrossing through texts and among fields of practice, to sort out their fuzzy resemblances, to appreciate distinctions of both/and rather than either/or. We might say that any media text is structured-in-dominance by several modalities in a mixed economy: while the mode of classical cinematic fictional narrative may dominate Renoir's film, the documentary mode's relative autonomy shifts the balance during the scene of the rabbit hunt to evoke the set of connotations we have come to associate with documentary's intention toward authenticity.

Returning to amateur practice, by adopting this dialogic notion of modality, we might better appreciate the interplay of conservative home mode moments in the work of Jonas Mekas and Stan Brakhage, whose films are generally structured-in-dominance by the avant-garde, without feeling obligated to conclude that their work is compromised or that home mode practitioners should try to emulate their avant-garde pursuits. As Stephen Neale has argued in his study of genre, "it becomes important, indeed essential, to differentiate between the various modalities of pressure involved, and to relate them to the various modalities of the political, ideological and economic conditions in which they function and take effect."[5]

This necessity for modal taxonomies is therefore one of pluralization rather than polarization—a method to analyze the different effects that competing cultural functions may precipitate within a media text and to prevent holding that text accountable to a single or "pure" intention.

Discussing the symbolic strategies of signifying practices, Sol Worth has argued that we should look for meaning not within the sign itself but in its social context, whose conventions dictate strategies of production and interpretation: "It is our knowledge of the conventions which govern social behavior in general, and communicative behavior in particular, that allows us to determine the intentionality of behavior, and hence the nature and extent of accountability that may be appropriate in a specific situation."[6] Worth denies that intention can be located as an empirical datum but posits instead that it can be derived from the assumptions that any symbolic sign-event has been structured intentionally for the purpose of implying meaning, and that its messages are probably intended for those who share its codes.

When informed by Bourdieu's concept of habitus, Worth's discussion of a sign-event's process of meaning attribution and interpretation assists our understanding of the symbolic strategies of the home mode. Unlike critics who advocate an amateur practice defined by phenomena worthy of investigation, Bourdieu contends that the field of photographic practice, in particular that subordinated to domestic functions, is not defined in relation to the object or subject photographed.[7] Dedicated not to the act of photography itself, home mode participants use cameras to acknowledge, communicate, and reconfirm the set of dispositions, codes, and conventions common to their habitus. Therefore, rather than deem the home mode practitioner's ignorance of a camera's complex functions and eager adoption of automated technologies as indicators of discursive conditioning by manufacturers, Bourdieu asserts that "it is the photographic intention itself which, by remaining subordinate to traditional functions, excludes the very idea of fully exploiting all the camera's possibilities, and which defines its own limits within the field of technical possibilities."[8]

By acknowledging different sets of intentions within the field of amateur practice, Bourdieu's notion of photographic intentionality understands how avant-garde and home modes express different cultural functions depending on the practitioner's symbolic purpose at the time of image making. If our habitus is structured-in-dominance by the home mode, we may tend to produce family representations. Or like Mekas, if our habitus is structured-in-dominance by the avant-garde, we may tend to produce alternative representations by exploiting complex techniques. But as illustrated by *Lost, Lost, Lost* (1976), a film in which Mekas attempts to construct a family within the New York avant-garde art community, affiliations with more than one habitus may overlap. After all, even the most conventional home videographer may experiment with an odd angle or unusual sound effect on occasion.

We might distinguish avant-garde and home mode intentionality by two sets of functional vectors: conservative/radical and aesthetic/social. The conservative vector functions as a mode of "articulation," which cognitive psychologist Arthur Applebee describes as practices that seek to preserve and legitimate a set of beliefs so that they may be passed on intact to a new generation. The radical vector functions as a mode of "reformulation," which Applebee describes as practices that challenge a system of values, seeking to extend its range or alter its basic principles.[9] Although the conservative and radical often align with right and left political agendas, we should not theorize articulation and reformulation according to this one-to-one correspondence. Instead articulation is better thought of as conservation, and reformulation as transformation. Therefore, within any community, the conservative vector wishes to preserve any set of values seeking continuity over time and space between generations, and the radical vector seeks to overturn any doctrines or dogmas operating as an oppressive set of ideologies within a community. This flexible system of relatively autonomous vectors prevents essentializing subjectivity according to the trends of political correctness at the same time that it allows for ideological critique of reactionary practice. Indeed, both vectors may intersect within a single sign-event: a videotape celebrating three generations of the women's movement is, for example, an articulation of the reformulation of patriarchal values.

The aesthetic vector functions within media artifacts that foreground form. Practitioners intending the aesthetic vector generally think of themselves as artists, and their products as works of art. The social vector functions within media artifacts that foreground content. Practitioners intending the social vector generally think of themselves as documentarists, and their products as communication. Once again, these vectors may intersect across a single text, which will usually be structured-in-dominance by one or the other. Thus whereas Brakhage predominantly intends the aesthetic vector in *Dog Star Man* (1961–1964) and Grierson the social vector in *Night Mail* (1936), Mekas alternates vectors in *Lost, Lost, Lost,* both documenting the avant-garde's community and demonstrating its visual codes.

As the example of Mekas illustrates, these functional vectors may intersect and share fuzzy resemblances within a particular media text. Alternately, by heuristically conceiving these vectors as a functional taxonomy, we may locate and understand their distinctive social intentions that may not be readily apparent simply by reading off their textual signifiers alone. Thus, within the field of amateur practice, we can broadly situate the avant-garde mode as structured-in-dominance by the intersection of the aesthetic and social vectors with the radical vector, and the home mode as structured-in-dominance by the intersection of the social and conservative vectors, without reducing one as essentially progressive and the other as essentially reactionary, but attending to the specificities of their functional values from text to text.

Within the avant-garde, we can distinguish two submodes. As the effect of the intersection of aesthetic and radical vectors, avant-garde formalism foregrounds gestures of rebellion against prevailing aesthetic conventions. Inventing shock tactics, methods of defamiliarization, and new audiovisual codes of perception, this submode of the avant-garde, represented by artists such as Nam June Paik, demystifies dominant media practices by foregrounding their materials and processes of production. As an effect of the intersection of social and radical vectors, avant-garde activism foregrounds gestures of rebellion against prevailing social conventions. Documenting oppressive living and working conditions, presenting arguments and demands, and stimulating action, this submode of the avant-garde, (formerly) represented by activists such as Michael Shamberg, demystifies dominant media practices by foregrounding alternative and self-determined representations of cultural and political identity, including race, class, and gender.

▶

The Home Mode and Functional Modality

Neither radical nor naive, the home mode displays a referential aesthetic that rejects photography as a disinterested practice and autonomous art. Defined in terms of use and users, this aesthetic is based on the subordination of form to function, of signifier to signified. Deeming images lacking a referent outside themselves as meaningless, home mode practitioners appreciate artifacts based on informative, tangible, and moral interest, judging their success according to whether they fulfill the intentions of participating members, from which home mode artifacts derive their purpose: "The feature common to all the popular arts is their subordination of artistic activity to socially regulated functions while the elaboration of 'pure' forms, generally considered the most noble, presupposes the disappearance of all functional characteristics and reference to practical or ethical goals."[10]

Rather than deriding the home mode for lacking an aesthetic in contrast to avant-garde formalism, we should see how it subordinates an expressive function to a referential function. In his important work on the home mode, Jean-Pierre Meunier has elaborated this referential function by outlining its phenomenology of spectatorship.[11] Discussing the *film-souvenir,* or home movie, Meunier distinguishes its mode of reception from other modes, such as fiction and documentary, by locating identification not with a mimetic image but with the absent person or event it signifies. Home movie representations, he argues, are taken up as evocations of things specifically known to the invested spectator, whose existence refers to a time and place beyond the confines of the screen itself. In this sense, a home movie's diegesis sutures on- and off-screen space with the shared life worlds of its participants.

Providing a sociological context for Meunier's phenomenological description, Bourdieu explains the home mode's referential function as an effect of the economic and cultural capital constituting the totality of its practitioner's habitus. Bordieu argues that financial and educational resources play equal parts in shaping one's tendency toward the home or avant-garde modes of practice: "The propensity to move towards the economically most risky positions, and above all the capacity to persist in them (a condition for all avant-garde undertakings which precede the demands of the market), even when they secure no short-term economic profit, seem to depend to a large extent on possession of substantial economic and social capital."[12] Bourdieu claims that any work of art has meaning and interest only for those who possess an understanding of its codes. The codes most meaningful to home mode practitioners are those that value content more than form and that uphold the presentation of subjects before the lens as more important than symbolic manipulations by the photographer behind it. These fundamental functional intentions distinguish the home mode's general governing aesthetic.

▶

Avant-Garde and Home Mode Autonomy: Problems of Accountability

By distinguishing avant-garde and home mode intentionality by variations in social function, material resources, cultural competence, and phenomenology of spectatorship, we understand their relative autonomy within the field of amateur practice as overdetermined by a variety of factors rather than merely as the effects of either increased public access and media literacy or deforming ideology. Nevertheless, as noted in chapter 2, by arguing that the home mode is a betrayal of amateur practice by dominant ideology, important critics have held home video accountable for what it is not, for what it ought to be: socially or aesthetically radical. This position can be traced in part to a series of problematic conflations of amateurism with the experimental cinema of Maya Deren, Stan Brakhage, and Jonas Mekas, whose avant-garde strategies and critical discourses flirt with features of home mode practice. Although structured-in-dominance by radical vectors, their films share techno-aesthetic similarities with home movies, a comparison that elides modal differences of social function in its focus on formal effects. By noting the textual affinities between, say, Brakhage's *Scenes from under Childhood* (1967–1970) and a conventional home movie, critics who champion experimental practice may be led into arguments by association for the potential in every amateur for the avant-garde. That is, if Brakhage can reformulate his obsessions with family and domestic life into a radical cinematic vision, why can't the average home moviemaker do the same?

As the leading pioneer of the "New American Cinema," Maya Deren pioneered the connection between amateurism and the avant-garde.[13] By setting the private, one-person form of film practice against corporate sponsorship and the production model of Hollywood studios, Deren implicitly marked the distinction between amateur and industrial not so much by economic relations as by her insistence on autotelism. Her definition of the amateur depended predominantly on her rejection of media practices organized by collaboration and division of labor, which, she argued, compromise the unity of personal vision by evaluating achievement according to a set of external standards and consensual goals. Tracing the etymological roots of the word "amateur" to the Latin for "lover," Deren declared that the amateur makes films for love and measures their success entirely by the work itself rather than by the recognition of others. Thus, in advocating an amateur practice autonomous from social and economic necessity and emphasizing unique personal expression, innovation, risk, and experimentation, Deren implicitly defined its aesthetic as avant-garde in nature, distanced from and transforming all established conventions.

If Deren conflated amateurism with the avant-garde, Stan Brakhage conflated both with the home mode. By appealing to the historical avant-garde's deliberate and critical intention to integrate the practice of life and the practice of art, Brakhage made filmmaking the agency of his being.[14] Therefore he upheld home movies as a practice integrating film and everyday life.[15] As David James tells it, after the theft of his 16 mm equipment, Brakhage turned to 8 mm, symbolically identified with home movies, which gradually served as a paradigm for his own aspirations.[16] Like home mode practitioners, Brakhage worked outside of industrial relations, but within the domestic sphere: "I am guided primarily in all my creative dimensions by the spirit of the home in which I am living, by my very own living room."[17] In films such as *Window Water Baby Moving* (1959) and *Thigh Line Lyre Triangular* (1961), Brakhage's dominant tropes of his wife and children serve various cultural functions of the home mode, including the documentation of events in the service of historical continuity. As P. Adams Sitney has observed, many of Brakhage's films act directly and solely to the needs of familial commemoration, in some cases appropriating photographic albums as a scaffold of memory and as a universal signifier of the amateur's desire for submergence of the self within a family.[18]

Where Brakhage reformulated the referential aesthetic of the home mode, Jonas Mekas reformulated the idea of home itself. Jeffrey Ruoff has noted that Mekas exploits the home movie's kinship associations to create a new home for an artist in exile, and has cataloged the cultural functions and conventions that Mekas's films share with home movies: an infusion of nostalgia; a tenuous link to the past, often closely tied to childhood; documentation of family and friends; quotidian subject matter dependent on contextual knowledge and familiarity with the people and events depicted; reliance on memory as an interpretive faculty; celebration of leisure activities; clowning for the camera; casual first-person voice-over narration recalling spoken commentary; use of locations rather than sets; abrupt changes in time and place; flash frames, in-camera editing, rapid camera movements, variable exposure, and jump cuts.[19]

From the avant-garde practices of figures such as Deren, Brakhage, and Mekas, all of whom appropriated home consumer equipment in the service of radical intentionality, whether to resist Hollywood paradigms, reject photographic realism, or seek cultural membership within a community alternative to the nuclear family, some media critics have deduced that the inherent value of amateur technologies is oppositional, and that, inversely, an amateur practice that fails to realize tactics of revolution must therefore be misguided. In summary, let us trace this itinerary. First, the category of amateurism, a strictly economic relation, is defined as a uniform mode of practice wholly constituted in opposition to a narrowed set of industrial technologies, ideologies, and aesthetic effects. Second, experimental artists, such as Maya Deren, reject the utilitarian standards of these dominant practices by privileging autonomous artifacts whose appreciation requires cultural competence, distance from social necessity, and the rejection of conventions—a modernist aesthetic lending itself more readily to reformulation than articulation. Third, artists such as Brakhage and Mekas, for whom amateur practice consists of eliminating the distinction between life and art, link the home mode, concerned with integrating media practice with the everyday, directly to the goals of the historic avant-garde. Fourth, by identifying the similarities between the visually inventive techniques of the avant-garde and stylistically "naive" techniques of the home mode, both modes are conflated by ignoring deeper functional differences and emphasizing superficial techno-aesthetic correspondences. Fifth, progressive media critics invested in advocating radical practice turn to these artists and to this narrow alignment of the avant-garde and home modes as "proof" of amateurism's revolutionary potential. Finally, condemning the home mode's conservative functions as betraying this potential, these critics hold its practitioners accountable to the intentions of the avant-garde and the ideological conditioning of false consciousness. Failing to see the avant-garde and home modes as relatively autonomous, both of which may surface within a single amateur artifact, they oppose the home mode as the avant-garde's antithesis.

This itinerary undergirds the utopian faith of progressive media discourse in a "pure" radical intentionality that will counteract all aspects of dominant ideology and practice. Its final conclusion not only disregards the cultural origins and social functions of the home mode but appeals to a set of myths about the future media "revolution" that have so far failed to materialize. In deconstructing the logic of these myths, my aim is not to deride avant-garde artists or their works but to interrogate the more problematic claims of critical discourse that the avant-garde should supplant less radical modes of practice because of its greater capacity for social transformation.

To begin with, the myth of art/life practice that has linked the avant-garde and home modes betrays a fundamental paradox that must be interrogated. As Peter Bürger reminds us:

> The avant-gardistes' attempt to reintegrate art into the life process is itself a profoundly contradictory endeavor. For the (relative) freedom of art vis-a-vis the praxis of life is at the same time the condition that must be fulfilled if there is to be a critical cognition of reality. An art no longer distinct from the praxis of life but wholly absorbed in it will lose the capacity to criticize it, along with its distance.[20]

Thus if the home mode may be condemned for constructing alienating ideal representations that misrecognize the real conditions of everyday life, any realization of the avant-garde's goal of art/life practice would prove equally alienating in its inevitable incapacity to recognize its own critical function. Does this choice of alternatives suggest, therefore, that both the avant-garde and home modes are doomed to alienation and duplicity? Only if we succumb to a theoretical dichotomy that opposes leisure/work and art/life as antithetical binaries.

Victor Turner proposes a more complex notion of leisure that revises its simplistic equation with distraction by Marxist fundamentalists.[21] Turner argues that leisure arises not from the alienation of labor but when the limits of what constitutes work become more arbitrary, when society ceases to govern its activities by means of common ritual obligations, causing some activities to become subject to individual choice. Providing freedom from institutional obligations and chronological, regulated rhythms to enter new symbolic worlds of play, leisure constructs a liminal sphere in which to explore personal identity.[22]

This view of leisure acknowledges a space for tactics to negotiate private and public life and to construct a more positive image of the everyday outside the confining strictures of work. We might argue, therefore, that the home mode's ideal representations signify not alienation but an alternative vision of the world that implies a critique of existing social formations.[23] In this sense, by their efforts to construct a utopian vision of the world, the avant-garde and home modes intersect in their imagination of social alternatives. Although less radical than the avant-garde, the home mode does nevertheless bear an implicit critical function, which counters conceptions of its practices as wholly expressive of dominant ideology.

A second myth of the avant-garde is that its practices are more "democratic" than those of the home mode. Like the myth of art/life practice, the myth of democratic practice suffers from an internal contradiction: if the activist community seeks to democratize media production, its effects are still electoral rather than genuinely collective. Like Michael Moore and his "TV Nation," one individual generally speaks for the rest. In part determined by the apparatus, whose eyepiece lends itself to the vision of an individual consciousness, typical activist films and videos tend to express a single perspective and sensibility. Dziga Vertov, often cited as the father of avant-garde media practice, affirmed the connection between the camera lens and the filmmaker's unique subjectivity: "I am kino-eye, I am a mechanical eye. I, a machine, am showing you a world, the likes of which only I can see."[24]

Jay Ruby has observed that the activist filmmaker's view of the world is paramount, even when his goal may be to document the lives of a community in which he participates:

> The possibility of feedback did cause the makers to think about the community impact of their work. However, it did not necessarily indicate a significant alteration of the relationship between the filmmaker and the filmed. The directors may have come from the communities they filmed but most continued the dominant pattern of maintaining control over the production of the film as an artist.[25]

An accumulation of privileged, atomized perspectives, the "collective" nature of radical practice seems more feudal than democratic, its noble intentions courting attention only among those willing and able to give the activist his or her due.

To argue that the home mode is simply a private, personal form of expression opposed to the public, collective expression of the activist mode must ignore the avant-garde's own history of idiosyncratic visions and electoral artifacts proxying for truly democratic media production, as well as the home mode's social function as a tool for community integration. Because they are personally involved behind and in front of the camera, and deeply invested during exhibition, home mode practitioners exercise a vital role in all aspects of production and reception, perhaps more so than in any other media practice available. Roger Odin writes:

> The institution of home movies produces a spectator who is more a "participant" than a real spectator: he takes part in the direction of the film (having held the camera), in the action taking place on screen (having been filmed), in the installation of the projection equipment (having set up the screen and projector), and, finally, in this type of event that consists of the collective creation of a memorial diegesis by the members of the family.[26]

Within the field of home mode practice, therefore, the range of roles available can be seen as more democratic than in the field of activist practice. Only when "collectivism," like "amateurism," is conflated with the avant-garde can the home mode be held accountable according to these terms.

Linked directly to arguments for radical democratic practice, often cited as its potential cause, is the myth of the mobilizing power of amateur technologies. This myth originates in a faith in the technological foundation of an alternative society, particularly of Marxists such as Brecht and Benjamin, who "tended toward fetishizing technique, science, and production in art, hoping that modern technologies could be used to build a socialist mass culture."[27] Like the inexpensive 16 mm film formats championed in the early twentieth century for providing the masses with greater access to media production, the advent of video in the sixties proliferated similar discourses of emancipation.

In particular, Hans Enzensberger upheld mobile video technologies for their liberation from the broadcast/receiver model of broadcast television and their potential to transform the masses from passive consumers to active producers both sending and receiving messages.[28] Not a technological determinist, Enzensberger traced the contradiction between production and consumption to economic and administrative institutions rather than to inherent properties of the media themselves.[29] Nevertheless, like his Frankfurt school forebears, he did believe that access to the means of production might be sufficient to begin the revolution.

In his article "The Myths of Video," Nicholas Garnham disputes the wishful thinking that access to the means of production empowers democratic media participation and social transformation. Instead he argues that video technologies cannot counter television's hegemony if their artifacts lack widespread distribution. Because television is the dominant medium of exhibition, control over representation rests in the structures of broadcast, not alternative subject matter. And because corporate financing and frequencies, rather than videotape, are the means of production that possess real effective power within culture, amateur video will in no way alter this situation.[30]

In summary, the myths of the media revolution that hold the home mode accountable to the avant-garde simply don't hold up to scrutiny. In particular, the notion that amateur media technology can be made to serve a just society or to inaugurate a counterculture simply by altering its uses must be rejected as mechanistic and empirically ungrounded. Again, my point is not to derail the goals of progressive media discourse for alternative practice but to deconstruct some of the premises criticizing the home mode for hindering radical intentions that may be paradoxical and unrealizable through *any* media practice.

In her analysis of alternative technology movements, Jennifer Slack reminds us that, ironically, technological interventions of progressive activists are generally vulnerable to exploitation by the corporate structures they despise.[31] Instead of focusing on technology as the origin of change, she advocates analysis of cultural determinations, such as the ideology of progress, regulatory practices, or the role of elites.[32] Following her cue, let us turn now to a detailed analysis of event videography as a case study demonstrating how the mobilizing power of video technology has, in fact, precipitated not a democratic art/life practice but an industry intervening in the field of the amateur to carve out a niche for commercial exploitation of home mode traditions. Marketing themselves as legitimate professionals in contradistinction to illegitimate amateurs, event videographers have appropriated the very equipment championed by activists—in the service, however, of commodification, not revolution—indicating that technological innovation alone cannot guarantee radical consciousness or intentionality. Yet even though avant-garde and event video may seem antithetical politically, in their concerted efforts to denigrate home video, sociologically they link arms as enterprises upholding elitist cultural distinctions.

▶

Event Videographers: Professional Amateurs or Amateur Professionals?

As objects of cultural analysis, commercial event videos fall outside the typical lines of theoretical inquiry. As media artifacts, they are neither broadcast nor publicly exhibited. As commodities, they are rarely mass-produced or advertised. Their producers cannot be located within specific, centralized institutions, and their audiences are always local at the most microdemographic levels. Less than cinema, more than a home movie, and aspiring to television, they complicate conceptual models of "mass media."

In the sense that they are both everywhere and nowhere (a research dilemma), event videos illustrate postmodern dispersal and implosions of classic binaries: private/public, amateur/professional, artisanal/industrial.[33] As simulations of home video, they hybridize production and reception, conflate autobiography and ethnography, and thus suggest unexplored configurations of image representation, self-identity, and memory. At the same time, as they increasingly infiltrate home mode practice, event videos demonstrate the continuing need for critical theory to situate their influence within the broader political economy and to examine the ways in which the instrumental values of professional videographers invest in, appropriate, and alter home mode conventions as they delimit and perhaps reinvent the genealogical traditions of family and domestic culture that, ironically, they have been hired to preserve.

An important aspect of event videography's power and prestige is its capacity to acquire and demonstrate the trappings of professionalism.[34] Primarily, the professional videographer must convince "laypeople" that they are unequipped, innately and technologically, to record their own ritual ceremonies, a practice better left to the more qualified craftsman who will better serve the general public's true interests. This claim to indispensable expertise, however, while appealing to professional ideology, lacks the conventional qualifications of "legitimate" professions, such as medicine, law, or education. Event videographers, for example, need not join associations that act as guardians of access to the profession; no code of ethics and licensing laws guide or control their practices; and most important, their credentials do not depend on institutionalized training or proof of specialized skill.[35]

Instead event videographers must arbitrarily and willfully construct their professional qualifications by discourse alone: "We all have a common goal to say 'hey look, we're not amateurs.' There's this stigma attached to event videographers, which isn't warranted. Anybody can pick up a video camera, but when you play a tape shot by me and one shot by a relative, I guarantee you'll see the difference."[36] This guarantee is, of course, more rhetorical than warranted, a fiction whose efficacy appeals to the videographer's reputation as often as it does to the actual audiovisual attributes of his productions.[37] Because home and event videos are generally produced with the same equipment and by practitioners with the same level of media education, the differences between amateur and professional artifacts are frequently negligible; therefore the event videographer must market himself as much as his commodities.[38] In short, the social status and cultural identity of the "professional" videographer, rather than the properties of his artifacts, mark the real distinction between event and home video practice. Because his "clients" (more accurately, "customers") may not recognize a significant difference between their practice and his own, he must rely on a set of cognitive codes that imbue his product with a professional aura transcending its empirical formal attributes.

One such code is expertise, which transmutes technical competence into social authority. Michel de Certeau has argued that the expert frequently confuses his authority with technical discourse, which makes him and others believe in the objectivity of his superior abilities.[39] In the field of event videography, this discourse emphasizes polished techniques and cutting-edge equipment as the "objective" standards supporting the professional's claim to monopolize practice. But as Geoff Esland has observed, a major threat to monopolies is the existence of other groups claiming similar benefits and offering different but apparently equivalent skills. When conflicts of this kind arise, they are often waged through campaigns of mutual denigration or through co-optation.[40] Home video therefore represents event video's greatest threat, particularly because it records events using similar equipment and techniques for free—its primary benefit. Thus the event videographer's campaign to legitimate his social authority requires not only affirmation of his competence but negation of the supposed incompetence of home videographers who may replace him.

Because its specificity as a media practice cannot be traced to techno-aesthetic norms proper only to itself, event video functions more as a cultural index. In his monumental study of photography, Bourdieu has demonstrated that only social distinctions can categorize its practices, which can be accessed by the vast majority, if not all, of the general population:

> It follows that the intention to give value to a practice as accessible as this necessarily includes at least a negative reference to ordinary practice. . . . In this way photography provides a privileged opportunity to observe the logic which may lead some members of the *petite bourgeoisie* to seek originality in a fervent photographic practice freed from its family functions.[41]

Here Bourdieu diagnoses one potential reason for the avant-gardists' dismissal of home mode practice: the need to mark themselves off as members of an elite class separate from the uneducated and commonplace masses. But if we apply Bourdieu's insight to the field of event video, we may observe an interesting twist: professional videographers distinguish themselves from home videographers as an alibi for economic exploitation—

and rarely for artistic expression. At once bourgeois in his practice of family photography and petit bourgeois in his efforts to seek autonomy from ordinary practice through expertise, the event videographer strives for originality, not necessarily in the form or content of his artifact but in his social status as a newfangled professional: who he becomes is more important than what he photographs.

Because both home and event videographers practice family photography, professionals must invent formal distinctions among artifacts and innate distinctions of skill among practitioners to defend distinctions of social authority:

> Explicit aesthetic choices are in fact often constituted in opposition to the choices of the groups closest in social space, with whom the competition is most direct and most immediate, and more precisely, no doubt, in relation to those choices most clearly marked by the intention (perceived as pretension) of marking distinction vis-a-vis lower groups.[42]

Thus, for example, the professional videographer frequently defines his techniques in opposition to those that, available to the amateur, are often not exploited by home video practitioners, such as using a tripod, artificial lighting, additional microphones, and editing decks. But because the event videographer understands that home videographers may also appropriate these techniques and equipment if so inclined, he must defend his skill as a matter not of choice but of natural disposition, taste, and talent. This discursive construction of the "born professional" thus protects the event videographer, by essentializing amateurs as "laymen at heart," from those who might imitate his ambitions.

Materializing these discourses of expertise and innate disposition within the field of economics and class relations, the event videographer further distinguishes his practice from home video by a set of invidious distinctions. Cashing in on his customers' desire to express conspicuous consumption as a marker of economic and cultural capital—of their ability to afford "the best" video services that money can buy—the professional tends to market the most elaborate productions, disguising his pursuit of maximum profit as the customer's sense of good taste. Thorstein Veblen writes:

> The consumption of expensive goods is meritorious, and the goods which contain an appreciable element of cost in excess of what goes to give them serviceability for their ostensible mechanical purpose are honorific. The marks of superfluous costliness in the goods are therefore marks of worth—of high efficiency for the indirect, invidious end to be served by their consumption; and conversely, goods are humulific, and therefore unattractive, if they show too thrifty an adaptation to the mechanical end sought and do not include a margin of expensiveness on which to rest a complacent invidious comparison.[43]

The event videographer's appeal to the discourse of professionalism supports the exploitation of invidious distinctions by constructing event video as "honorific" and home video as "humulific," a system of opposing values determined more by quantity than by quality, more by the professional's power to increase his commission than by his capacity to ensure excellence.[44] We see, therefore, how acts of consumption, more so than methods of production, mark the distinction between home and event video: because no internal attributes necessarily distinguish them, the event video's professional value can be consummated only by its sale on the market. The event video's material signifiers refer not to any value inherent to it as an artifact but to a completely arbitrary discourse buttressing the producer's ability to charge and encouraging the consumer's willingness to pay for what remains home video at base.

At the same time that event videographers exploit their customers' desire for invidious, conspicuous consumption that emulates mobile class aspirations, they also seek for themselves a reputation that emulates corporate industrial agency. Although they market their videos to the general public as professional in nature, their efforts to sustain this authority are constantly undermined by the standards of commercial television that define their production methods as amateur. For example, the popular TV series *America's Funniest Home Videos* disqualifies industrial video from competition but regularly airs home and event video as equivalent. How embarrassing for the professional wedding videographer trying to build a reputable business whose "expert" work is broadcast alongside Uncle Charlie's!

Stranded between amateur methods and professional aspirations, event videography illustrates what Bourdieu has called "semi-bourgeois careers."[45] Found in the newest, most ill-defined and unstructured sectors of cultural and artistic production, these occupations invent new positions to prevent downclassing. Because their fields of practice are indeterminate, not yet acquiring the rigidity of older bureaucratic professions, they offer the most favorable ground for operations that aim to produce spheres of specialist expertise based on interpersonal connections and a rationalized form of competence rather than genuine qualifications. Positioned halfway between studenthood and a profession, the semibourgeois occupation provides its practitioners a provisional future that makes it possible to view their present status as endlessly renewable, avoiding or refusing limits to their imagined aspirations. Many event videographers discuss their decision to pursue this new industry in exactly these terms: to escape the confines of nine-to-five employment, to structure their work and leisure time according to their own dictates, to imagine prosperity limited only to their self-determination and marketing skills, and most important, to cherish their new identities as "entrepreneurs" rather than "employees."

Whether event videography actually returns greater profits than a traditional paycheck seems to matter less than the occupation's accrued symbolic capital. Coined by Bourdieu as the third form of capital (along with economic and cultural) determining operations and transformations within a field of practice, symbolic capital measures the degree of accumulated prestige, celebrity, and consecration of honor founded on a dialectic of knowledge and recognition.[46] In this sense, event video represents a symbol of expertise and autonomy within the field of home mode practice even though it betrays most of the characteristics of home video. To convert cultural capital (knowledge of video technique) into economic capital (profit), event videography therefore depends on symbolic capital (professionalism) as its primary conversion strategy.

One method by which event videographers seek symbolic capital is by establishing formal video associations. *Videography* magazine's "Professional Special-Event Videographer's Handbook" cites the two primary goals of the Society of Professional Videographers as "public awareness and professional enrichment. Since the nonprofessional stigma of this market is a major deterrent to its business opportunities, market maturity is the ultimate goal of the SPV."[47] In the same article, wedding videographer Randy Kinsley of RDR Video Productions in Northbrook, Illinois, affirms that even though he had been making good money for years, he joined an association because he failed to get his due among professionals: "It's tough sometimes because organizations like ITVA and NAB don't recognize the wedding market."[48] Although the credentials for joining these associations generally require only proof of commercial activity and payment of annual dues, membership attempts to imitate the symbolic cachet of reputable professional organizations and hence justify exploitation of the home video market.

A second conversion strategy concerns the invention and adoption of "prosumer" video equipment. Basically there are two grades of video technology. In the field of the amateur, "consumer" generally describes relatively inexpensive, lightweight, portable products available for purchase at commercial retail outlets. In the field of the industrial, "professional" generally describes relatively expensive, heavy, stationary machinery designed expressly for big-budget studio productions. Event videographers, who usually operate out of domestic spaces or small offices, use consumer equipment, therefore falling outside of the professional category they seek to join. Even more disconcerting, the colloquial term for consumer equipment is "home video," which locates their practice within the home mode that they seek to relinquish. The solution to this dilemma? "Prosumer" equipment: consumer video technologies boasting more expensive, more complex features and operations usually disregarded by home videographers as inessential, such as graphics generators, directional microphones, and additional chips. Lobbying video manufacturers extensively for the invention and sale of prosumer equipment, event videographers have been able to adopt material signifiers of their symbolic capital as middle-level professionals, converting their status as amateurs by distinctions of technology itself. Patricia Zimmermann has observed the same processes of conversion in her study of amateur film: "The distinctions between professional and amateur filmmaking equipment resided in the degree of technical control over exposure, focusing, and effects. Manipulation of technology, higher cost, and technical complexity denoted professionalism, and conversely, ease of operation, lower cost, and simplicity defined amateurism."[49]

As a hybrid adopted self-consciously by "professionals" to effect a transverse movement from home to event video, prosumer equipment serves as their "objective" standard of quality. Danny O'Keefe, of Dancel Productions in New Orleans, maintains: "I'm using high-quality prosumer equipment, not something I bought at a department store. So my product will be infinitely better."[50] Cloaking the event videographer's social ambitions in technological garb, prosumer equipment also enables a vertical movement in class mobility. Notes Zimmermann: "As ability improved, consumers could 'trade up' to a higher status by purchasing more expensive and complicated equipment. On the social and ideological level, this hierarchy of camera designs positioned filmmaking expertise as a passport to further professionalization of leisure usage with more and more expensive equipment and gadgets."[51]

This idea of trading up is wonderfully illustrated by *Videography* magazine's feature on Frank Farell, of Custom Video Production in Rumson, New Jersey. His professional biography reads like a rite of passage through stages of increasing technological sophistication, worth quoting at length:

> You could say that Frank Farrell was born with a videotape camera in his hands, as his first experience with video came in grade school, experimenting with 1/2" b/w reel-to-reel EIAJ videotape. In high school he got his hands on his first color camera and the rest is a success story that is still being written.
>
> His first "serious" video camera was a Sony Betamax camcorder, which he used to record events in his hometown. From there he moved up to a Panasonic PV-3000 VHS camcorder, then to a Panasonic D-5000 and an AG-6400 recorder. Then he upgraded his equipment to a Sony DXC-3000AK three-chip camera and an AG-7400A S-VHS recorder, and now he has a Sony

M7 three-chip broadcast camera with a Hitachi VLS-1000 S-VHS recorder. He also has a VO-8800 3/4" SP camcorder, which he uses for selected jobs. Farrell attended a branch of the Sony Technical Institute in New Jersey and has taught videography seminars at several colleges in New Jersey.

Throughout his professional career Farrell has been shooting an even mix of event videography and corporate industrial work. Although his work has appeared on many of the major news networks and on ESPN—doing both ENG and aerial photography—Farrell still feels he has to fight to be considered among the big boys. "It's a tough business because a lot of the stigma involved concerns the type of equipment being used. It's like who has the better car. A lot of event videographers are using consumer equipment so they're looked down upon. But having all the right equipment still doesn't make you a professional. "Professional" is two things: it's your tools and your talent. Talent is acquired over a long period of time, you can't learn this stuff overnight."[52]

Farrell's professional reconversion strategies depend on two distinctions: one, that there is a fundamental difference between the effects of consumer and prosumer technologies on practice; and two, that even if other amateurs appropriate prosumer equipment, they may be retarded by an immature professional disposition. Thus because the event videographer cannot restrict the home videographer's access to his field of practice through corporate and institutional checks, educational requirements, or licensing, he must align his practice with technologies boasting operations exceeding the resources and needs of most home videographers; and if this barrier fails, he may always resort to the discourse of talent, a more impenetrable (because ideologically naturalized) line of defense.

A third strategy of reconversion, the denigration of amateurs as "laymen," defines the event videographer's legitimacy by negating the home videographer's competence.

> A major characteristic of the professional mandate is that it is dependent for its justification on a somewhat negative view of the lay public. The clearer the boundary between professional and non-professional areas of operation, the more likely it is that there will be a set of assumptions concerning the means of controlling the lay public. The concept of the "lay public" is an important one in that it legitimates a profession's *raison d'etre*—that is, that it has a right to do things for and to people.[53]

One technique for keeping clients in their place and minimizing their intrusion into the professional's field of practice is to condescend to home videographers as well-meaning but laughably inept greenhorns. For example, event videographers often deign home videography as the work of "Uncle Charlie," a friend or relative invited to record a special family event as a favor or gift.[54] Because Uncle Charlie's intimate familiarity with the event's participants would make him more qualified than the estranged professional, at least in terms of home video's referential aesthetic that emphasizes recognition of appropriate faces at just the right moments, professionals must turn to criticizing his technical "errors," such as shaky camera movement, underexposed visuals, and audio dropout. Given the customer's choice between a confidant who knows the community involved, but who may obscure their representation, and a stranger unfamiliar with the community's social nuances, but who promises clear audiovisual reproduction, the professional highlights technical competence as the most important consideration in videotaping special events. By directing attention to Uncle Charlie's dubious skills, the event videographer evades the need to establish his professional credentials (or confess his lack thereof) and provides the consumer with an illusion of free choice between two fundamentally distinct practices.

Of course, because the event videographer uses consumer equipment himself or its prosumer variation, his emulation of professionalism is itself a simulation, betraying his lack of an authentically autonomous voice or reputation independent of home video's negation. This process of negation functions itself as a reconversion strategy disguising event video's purely economic distinction from home video as a distinction of techno-aesthetics. Because the event videographer, as a pseudoprofessional, must distract attention away from the probability that his status is authorized only by his temerity to charge for his services, he must justify his price by demonstrating that his customers' capital will translate in the product itself as visible evidence of the value of their investment—even though home and event videos often present no visible differences between them.

Event videos thus pose a challenge to hermeneutic methodologies that attempt to read off from the formal properties of media artifacts their mode of economic production. For example, in his study of avant-garde film, David James claims that "a film's images and sounds never fail to tell the story of how and why they were produced—the story of the mode of their production."[55] Coining the phrase "allegories of cinema," James argues that every film internalizes the conditions of its production, which are visibly encoded in the text itself:

> As the site of conflict or arbitration between alternative productive possibilities, they invite an allegorical reading in which a given filmic trope—a camera style or an editing pattern—is understood as the trace of a social practice. In such a materialist version of the Formalist reading of texts as the mobilization of preexistent devices, formal motifs signify through the text to the material practices that produce it.[56]

In particular, James argues that industrial modes of production may be identified in opposition to the social and formal alternatives of nonindustrial practices that are inscribed in the artifact as marks of difference from dominant discourse.[57]

It would seem, however, that some media texts are more allegorical than others. Brakhage's films, for example, would lend themselves to this mode of reading. But what of event video? Although its industrial nature embroils its practice in the instrumental values of commodification, its formal attributes and familial ideologies may nevertheless appear identical to home video, an amateur practice autonomous from the market. Therefore, as an industrial practice that often as not remains indistinguishable technologically, aesthetically, and ideologically from the home video it simulates, event video undermines the promise of materialist formal readings. Perhaps this approach finds better applications in the field of cinema than in video. DeeDee Halleck observes:

> With home movies, we always knew they weren't MGM. The size of the image, the clarity of registration, the skillful use of lighting all made "the movies" look very different from home movies. With video that differential is less evident. The new video cameras obviate the use of lights, and the resolution of detail is approaching that of some network programming. So too the use of a hand-held style, casually zooming in and out, has appeared in IBM commercials and on MTV. All this has brought a meeting of the images that puts in doubt any attempt to hierarchize "professional" video photography.[58]

To understand the significant distinctions between home video and its commercial simulation by event videographers, we must turn away from internal markers to the social and cultural contexts comprising their fields of practice, in particular issues of consumption, intention, and cultural function. According to Michele Barrett:

> Meaning is not immanent; it is constructed in the consumption of the work. Hence, no text can be inherently progressive or reactionary; it becomes so in the act of consumption. There may be an authorial "preferred reading" but the effects may be different from, even opposite to, those intended. Whatever the formal properties of a work, its ideological content, its "political" implications, are not *given*. They depend upon the construction that takes place at the level of consumption.[59]

Taking her cue, for the remainder of this chapter, I will explore a few of the broader issues of intention and consumption raised by event video's simulation and commodification of the home mode's cultural functions. First, I contextualize event video as a mode of production inflecting affective relations with instrumental values extrinsic to the home mode. Second,

I analyze professional wedding videos to interrogate the ways in which event video complicates and commodifies home video's potential for inter-subjectivity. Third, and finally, I analyze the recent phenomenon of memorial videos to critique how event video compromises and co-opts the home mode's ritual intentions toward self-determination.

▶ ───

Event Video: For Better or for Worse?

In *Capitalism and Communication,* Nicholas Garnham characterizes late capital's search for new markets as a development of the service sector driven to industrialize sectors that have been either more primitively organized or altogether outside the market in the sphere of domestic labor. Furthermore, Garnham notes that in capital's attempt to increase productivity in sectors resistant to such increases, the struggle within cultural fields of production must be sharpened by new methods.[60] In short, late capital's commercial interests have come to dominate areas of life that were until recently dominated by individuals themselves.

Fitting Garnham's profile, the event videographer infiltrates the amateur field, professionalizes his own practice to construct a demand for his services, and simulates the communal ties of the home mode paradoxically by appropriating the detached codes of technical expertise. Because professional-client relations are generally estranged, whereas kinship, family, and friendship ties are affective, in order for the event videographer to simulate home mode intentionality properly, he must willfully stimulate an emotional connection to the community of participants he has been engaged to record. Expected to perform in theory and practice as an implied but anonymous witness to the event he is documenting, he must conceive of his medium as a representation more expositive than interpretive; he must adhere to rather than question the values and symbols being celebrated. More like "professional amateurs," event videographers mimic the amateur's ideological perspectives, but with supposedly expert techniques.[61]

Professional wedding videos provide excellent examples of some of the problems resulting from the industrialization of the home mode's autobiographical functions. As a hired "ghostwriter" of sorts, the wedding videographer both complicates and commodifies the home mode's intersubjective relations, first by negotiating a complex set of identification strategies to present the ceremony from the bride's perspective rather than his own, and second by entangling the wedding video itself within the instrumental values of the market, which convert a private experience into a public commodity and circumscribe the bride's agency as producer of her own story. Thus on the one hand, the practice of event videography itself, although professionalized, may hold out potential for a genuine collaboration between bride and videographer. On the other, this collaboration may be fraught with power relations endemic to event videography's commercial nature.

The complex relationship of autobiography to the professional wedding video is wonderfully illustrated by a cover story in *Videomaker* magazine that offered professional videographers "pointers" for recording successful wedding videos, accompanied by a photograph of a bride in full regalia peering through the viewfinder behind a camcorder. Were the editors here suggesting that the bride videotape her own wedding? Making a more metaphorical point, what the photo suggests is that a wedding video should be construed as the bride's own story. Although technically a biography in production, it should function as an autobiography during reception.

As a social and technological arrangement, the wedding video production is a hybrid collaborative form that both facilitates and conflates autobiographical and ethnographical expression. This partnership of bride and videographer sustains the common ideology of the home mode that the professional wedding video tries to simulate: an attitude that the person behind the camera has little to do with the photographic process, that "the presentation of oneself and manipulation of oneself are *more* important than controlling and manipulating the symbolic content from behind the camera."[62] Rather than merely posing his subject, the wedding videographer speaks with, not for, the bride and thus complicates the division of self and other in typical ethnographies. As the subject of vision and object of representation, the bride constitutes the classic autobiographical divided self who regards herself in the mirror of her own prose. Video lends itself even more to this doubling, because the bride may see herself in a monitor while her recording is being generated. Here enters the videographer, hired to do the "writing" for her.

Sharing rather than usurping the site of representation, videographer as subject meets bride as subject in a self-conscious awareness of self *with* the other rather than *against* her. In planning and coordinating the wedding herself, providing the videographer a shot list, contributing photographs, music, and creative ideas for the video, or reviewing the rough cut for criticism prior to final copy, the bride may exert a considerable degree of control over her own representation. Her image as "bride" is neither hers nor the videographer's, but a shared projection of both.

In theory, the integrity of the home mode's intersubjective relations may potentially be preserved during production. However, because event video stresses expertise and product over personality and process, its professional and commercial nature compromises true intersubjectivity by infusing social relations with relations of power: "For a production to be truly collaborative the parties involved must be equal in their competencies or have achieved an equitable division of labor. Involvement in the decision-making process must occur at all significant junctures."[63] The mechanics of event videography generally tip the scales: the videographer is granted final authority as an active agent, and the bride is reduced to a passive source. Entrusted with editing the wedding video, the videographer chooses what is relevant, not according to the unique identities and idiosyncratic relationships among the event's participants, with whom he is barely familiar, but to the generic conventions of marriage ceremonies relied upon to tell a coherent story, and to the laws of the market that will guarantee the video's success as a commodity.

The capacity of the event videographer to share the process of representation with his client fails to efface the gendered structural inequalities of the political economy in which the wedding video circulates. At bottom line, the videographer's investment in the bride concerns her cash, of course, not her consciousness, thereby infusing her subjectivity, memory, and pleasure with instrumental values.[64] The intersubjective exchange of self and other during production is reconfigured in the realm of consumption as an exchange of image and capital, of memories for a fee. Depending on the bride's economic status, therefore, the quality of her video will be determined far less by the investment of her creative agency than by her capacity to afford designer options within her financial means. The difference, we might say, between a two- and three-camera setup, between cabled and wireless mikes, between in-camera and postproduction effects, is also a class difference, whose economic exigencies prescribe the technological resources employed and mark the limits of the video's capacity as self-portrait. The clear exchange of vows on audio, for instance, or of rings in close-up, valued for their religious and private sentiments, become subject more to the dollar sign than the sign of the cross, to commercial rather than to interpersonal exchange.

Displayed to prospective clients in his studio and at bridal shows as a demonstration of the videographer's skill, the bride's personal wedding video becomes an alibi for male professionalism. Exploiting the bride's image and autobiography in the service of public enterprise, *her* self-portrait becomes *his:* a mirror of the videographer's creativity and a vehicle for his self-promotion. While the videographer advances in the "masculine" world of agency and production, the bride's empowerment is confined to the "feminine" realm of subjectivity and representation. In short, the wedding video empowers women only to a degree that cannot escape containment in a patriarchal economy that writes her power and denies her agency. The industry's ideology of male professionalism continues to engender myths of female-related technical incompetence to regulate access to the means of production, and to position women as consumers in the sphere of business relations. Stamping a price tag on the value of her wedding day, it is the male videographer, in the final determination, who regulates the bride's memories and her display.

Like wedding videos, memorial videos of the dead illustrate how event videography attempts to cash in on the human life cycle's every rite of passage. What makes memorial videos most significant is that although death seems to have been a patterned elimination of the home mode during the era of home movies, late capital's pressure for new markets has broken new ground for exploitation by video. Like marriage, death ceremonies may become increasingly subject to the conventional and generic interventions of event videography, which arrogates mourning rituals to the domain of professional functionaries and transforms the bereaved into passive consumers of premade ritual artifacts.

Both event videography and mortuary management share a similar history of pseudoprofessional aspirations: "The rise of commercialism in the funeral business began as death became more removed from the home. Nevertheless, the emergence of the funeral director is but one *instance* of a larger trend of *allocating functions* previously performed by intimates to officials in formal organizations."[65] The primarily instrumental nature of these industries has been misconstrued as professional authority and public service. Like the event videographer,

> the funeral director, who simply is not comparable to other professionals in the realm of professional authority, is similar to business people who know their products and services well and can assist customers in making selections efficiently. Yet, because of the mystifications surrounding death, the public aversion to both death and funerals, and the vulnerability of the bereaved, the funeral director is in an intriguing position by which trade information can be stretched into professional authority.[66]

Robert Habenstein has noted that during the development of the funeral service industry from 1850 to 1960, undertakers emerged as entrepreneurs whose secular mortuary tasks were self-consciously carried out within the context of business, but whose image appealed to the discourse of professionalism to disguise the purely commercial nature of their enterprise.[67] Perhaps the history of the mortuary industry's gradual evolution as a professional institution portends the future of event videography itself. In any case, because both services attempt to direct private rituals for display and profit, their partnership in the recent phenomenon of memorial videos makes good business sense.

One of the most successful video memorial companies, the National Music Service Corporation of Spokane, Washington, produces six-minute video tributes, which consist of a montage of photos of the deceased (while alive) accompanied by generic scenic backgrounds, programmatic music, and conventional religious or secular sentiments. This national service is offered to families when they are making arrangements for a funeral. Given the speed of express mail and the relatively uncomplicated format of the tribute, the finished videotape can be supplied within a few days to be shown at visitations, as well as during funeral, prayer, and memorial services.[68] At a more local level, Crawford Memorials of Duluth, Minnesota, markets video diaries featuring images and sounds that reflect the values of the deceased: photos, home movies, home video, graphic designs, lyrics, poems, letters, quotations, anecdotes, music, and voice-overs selected by family and friends. Marketed as an intimate coproduction, the videos are then dubbed and sold back to the mourners to take home.[69] Finally, John Dilks of Creative Tombstones in Pleasantville, New Jersey, has proposed a "talking tombstone" with a built-in video panel. The screen would be rigged with a "proximity detector" capable of picking up ultrasonic sound waves produced by the approach of a visitor to the grave. Once activated, the screen would transmit a series of synthesized images: the name of the deceased, his photographic portrait, a family tree listing ancestors and surviving relatives, and a digital clock giving the exact time of death. At the same time, a computerized voice based on that of the deceased would be emitted from the stone, introducing itself and launching into the deceased's life story.[70]

Despite their differences, each video memorial service markets primarily the same benefits: the commemoration of the life of the deceased, a legacy to be passed on to future generations, a document for family members unable to attend the funeral, the occasion for personalized service, and the opportunity to work through and express feelings about the deceased and to grieve in the privacy of home. Thus we can situate the memorial video as a hybrid of home video, in its domestic, private destination; of funeral management, in its public display and therapeutic function; and of event video, in its professionalized production as an audiovisual commodity.

Yet rather than confront real death, memorial videos may actually counteract their intended therapeutic benefits by reinforcing contemporary notions of the corpse as a cultural taboo.[71] In contrast to postmortem photography, video memorials are constituted by a fundamental structuring absence: the deceased as corpse is never represented. While few twentieth-century Americans seem offended by remembering the dead in videos taken while they were alive, postmortem remembrances are viewed more negatively. People who wish to record dead bodies face a personal conflict and potential public disapproval, trapped between two antithetical cultural norms: the use of video to memorialize important rites of passage and the belief that material reminders of death are pathological and anti-social. As Richard Chalfen's research on snapshots and home movies has demonstrated, conventional home mode practices generally ignored the physical process of dying, the precise moment of death, funeral ceremonies, and graveside interment. Conceived as bad fortune, personal failure, and loss of status, death obliterates the value system that the home mode emphasizes and strives to preserve: happiness, family togetherness, and success marked by the gradual accumulation of wealth, status, and goods. As an expression of personal progress, home mode images generally acknowledge visual transformations with a positive outcome; as an indexical sign less of death and more of the dead, the corpse signifies not a process but a thing devoid of semiotic power.[72] It disturbs our cultural and religious beliefs by denying that all rites of passage can be photographed, understood, successfully endured, and eventually overcome.

Sharing these characteristics of the home mode, memorial videos sustain the funeral industry's euphemistic, cosmetic effacement of death, rivaling, if not replacing, the "loving memory picture" of the made-up, dressed-up remains reclining peacefully in the "slumber chamber." This photographic repression of the dead is a reflection of contemporary disposal practices: the majority of deaths now occur in hospitals, where the fiction of probable recovery is often maintained until the patient is near the point of death. The corpse is then promptly removed without the aid of the bereaved, who see it again only under very special circumstances, if at all, after it has been embalmed to appear as if sleeping.[73] If the death is represented as real, a forbidden visual taboo is violated. Indeed, those who possess postmortem photographs regard them as very private pictures to be shown only to selected people, never pasted into family albums, but deposited in a special place of more limited access, like the erotica consumed in middle-class homes.[74] Like sex, death has become pornographic, a subject too controversial for public discourse and severely limited in representation.

More so than memorial videos, postmortem photographs seem to have greater potential to provide significant assistance to mourners by forcing them to accept the finality of the loss and begin their essential reintegration into society. Indeed, there has always been a commonsense link between grieving and the photographic arts. According to André Bazin, "If the plastic arts were put under psychoanalysis, the practice of embalming the dead might turn out to be a fundamental factor in their creation."[75] Describing the recording arts as a "mummy complex," "the preservation of life by a representation of life," the prevention of a "second spiritual death," and a rescue from time's "proper corruption," Bazin enlightens our understanding of how mourners may confront postmortem images with two seemingly contradictory needs: to accept the reality of loss, and to keep the memory of the deceased alive.

I would suggest that video memorials provide only the latter function. The very properties of still photographs—the irreversible suspension of time, the violence to our sense of duration, the conversion of subject to object, the displacement from this world into an otherworld of disembodiment, the incapacity to speak and return a look—these qualities still photography shares with death, are in effect reversed, time-shifted by video.[76] Whereas photography arrests, video reanimates. Video's capacity for the instant replay, pathologized as a repetition compulsion, reconceives of life as something that can be rewound, indefinitely deferring the inevitable loss of signals at the end of the tape. Video's implicitly naturalistic aesthetic, its ideologies of liveness, presence, and synchronicity, combat the ideologies of stasis, absence, and disjuncture. Video represents the success of a communications technology to keep the human being alive, despite the failure of medical technology to do the same. And if death separates individuals from the world of possessions and leisure, threatening our capitalistic values, the videocassette reunites them with the material world, as a commodity that can be owned and operated by friends and family within the safe haven of the domestic sphere.

Video memorials reprivatize and may potentially defer healthy mourning. Joan Crawford, of Crawford Memorials, advocates them as "a way to let families get home quickly to do their own private grieving."[77] This marketing ploy fits well in a culture increasingly lacking public support for grieving, leaving all such tasks to individuals who, in the postfuneral period, most urgently need community assistance with the distressing pressures of bereavement. Instead industrial memorial videos transform grief into a private burden that the funeral and event videography industries invest in to accumulate financial profits rather than to administer psychological healing.

As home mode practitioners increasingly hand over the management of life's rites of passage to professionals, they opt for custom-made ritual artifacts whose capacity for authentic reenactment usually resorts to hollow cliché. Joan Crawford, for example, in contrasting her personalized videos with the assembly line videos of the National Music Company, distinguishes their various services as "the difference between shopping at a boutique and at Wal-Mart."[78] While her sentiments may be admirable in their critique of services capitalizing on death rituals in their most impoverished, generic forms, they are also revealing in her metaphor of consumption. Buying a video memorial rather than producing your own is not the issue, so it seems, as long as you can afford death above bargain-basement prices.

If progressive media discourses wish to hold less-than-radical practices accountable, perhaps they should shift their focus from home video to event video, since its consumer imperatives truly transform amateur modes of production into acts of consumption. As a simulation of both amateurism and professionalism, event videography deconstructs the conventional (and increasingly complacent) theoretical dichotomies distinguishing industrial and nonindustrial media. Rather than antithetical enemies, modes of amateur practice might better be conceived as sibling rivals, whose features share family resemblances of technique and ideology, but whose distinct aspirations compete for the divided attentions of various critics and practitioners. If home video suffers the shame of the bastard's illegitimate status, event video returns with a vengeance as the prodigal son.

[4] *Family Resemblances: The Home Mode as Chronotope*

Video and television inhabit "relative" relations. Like fraternal twins who look alike but are not identical, one medium frequently is taken for the other. Residing side by side in domestic space and borrowing each other's technological garb, video and TV share so many likenesses that we recognize them as sibling media. Particularly at home, where they have been adopted to portray everyday life in amateur productions and network programming, video and television mediate, and are themselves mediated by, notions of family: while we use these media audiovisually to represent family relations to ourselves, we also use family relations discursively to represent these media to each other.

Ludwig Wittgenstein suggested the idea of "family resemblances" in his philosophical investigations to describe the existence of concepts that depend on the presence of their incomplete and overlapping similarities to be comprehended.[1] For instance, we understand "brother" as similar to "sister" in their filial relation to "parents," yet as different in their sexual relations to "mother" and "father." Although more concerned with the use of everyday language games, Wittgenstein's idea may be applied to the use of everyday media, such as home video and domestic television, and provides a useful analogy for the discourse of medium specificity in general. Two components of this analogy in particular prove most relevant to this study. First is the notion that any object category is graded, its members related to one another without all members sharing all of the properties in common that define the category (e.g., video and TV as electronic media). Second is the discovery that the cognitive aspects of set theory by which we construct object categories can trace roots to kinship ties as a trope for conceiving the relation among the category's members (e.g., video and TV as sibling media).[2]

Twentieth-century media discourse, from film to computers, has been no stranger to taxonomies ordered by prototypical kin relations. For example, cinema's earliest theorists relied on metaphors of family as one method for making sense of the new medium's relationships to its predecessor and contemporary art forms. In tracing the history of cinema to the nineteenth-century novel, Eisenstein declared how "it is always pleasing to recognize again the fact that our cinema is not altogether without parents and without a pedigree, without a past, without the traditions and rich cultural heritage of past epochs."[3] And Rudolf Arnheim, analyzing pure and impure hybrid media art forms, appealed to miscegenation analogies: "Their combination resembles a successful marriage, where similarity and adaptation make for unity but where the personality of the two partners remains intact, nevertheless. It does *not* resemble the child that springs from such a marriage, in whom both components are inseparably mixed."[4] As television developed from the thirties through the eighties, Eric Barnouw humanized the new medium in his standard historical survey. Imagining the evolution of TV as imitating the human life cycle, his chapter titles echo the order of a family album: *Forebears, Toddler, Prime,* and *Elder.* Note, for example, the anthropomorphic diction in his description of the toddler phase: "These were the formative years for television. Its program patterns, business practices, and institutions were being shaped. Evolving from a radio industry born under military influence and reared by big business, it now entered an adolescence traumatized by phobias."[5]

Never neutral, kinship analogies bring with them a history of their connotations that may inflect definitions of specificity with the power dynamics inhering in the familial relations with which the medium is compared. For example, in defining television against cinema, many media scholars have adopted gender metaphors to establish the fundamental distinction between each medium, citing certain characteristics of film as "masculine" and those of TV as "feminine." Film, for instance, has more frequently been analyzed within the contexts of production, active spectatorship, male voyeurism, and public entertainment, whereas television has been analyzed within the contexts of consumption, passive spectatorship, female fantasy, and private experience. Filmic metaphors often appeal to the eye and the paternal phallus, TV metaphors to the ear and maternal flow.

The primary problems with the masculine/feminine analogy are threefold: one, it essentializes masculine and feminine characteristics as antithetical; two, it sets film and television against each other as polar opposites; and three, it participates in a history of sexist aesthetic criticism that implicitly denigrates "feminine" forms, stigmatizing television as inferior to cinema. Not merely betraying the theorist's implicit biases about cultural production, hierarchical metaphors indicate that thinking of medium specificity in terms of family resemblances may lead to essentializing an entire medium according to one privileged (and often naturalized) connotation, transforming the prolific notion of resemblances into the either/or restrictions of dichotomy.

When we turn to an analysis of video, we see how significantly the politics of family resemblances exert their influence on defining the new medium's legitimacy, particularly in video's relation to its "parent" forms. Like cinema and television before it, video has been domesticated by anthropomorphic metaphors. For example, video's interface compatibility with other media has been imagined as sexual intercourse: "Having shaken many of the technical bugs out of its bed in recent years, the video industry is now deeply involved in an affair with 'an older man,' namely film techniques. . . . The two make a beautiful couple."[6] Likewise, video's appearance on the market has been described in terms of pregnancy: "Kodak's January 1984 announcement of the impending birth did nothing to slow half-inch VCR sales during the gestation period or beyond."[7] And once "born," video has been compared to a growing child: "Your video equipment may never need a new bike or braces or tuition for college, but from the moment you carry it out of the store, it does need a parent's kind of care and protection."[8] Indeed, almost like a poetic conceit, familial metaphors for video have often been pushed to exhaustion:

> Kodak and GE claimed paternity to 8mm video in January 1984, and delivered the diminutive format a year ago this month. Until now their offspring has been troubled by growth pains. All that seems about to change. Just in time for Christmas, at least a dozen other manufacturers will have produced siblings for the first-born pair. This new generation comes into the world better endowed.[9]

How do we explain this proliferation of kinship analogies? Roy Armes has suggested that at its origin, video lacked an obvious, unique purpose: "If a film is an artefact which aspires to (and indeed on occasion becomes) art, and original television strives to become an event, video is perhaps defined as a recording material in search of a mode of production."[10] For many media analysts, the speed at which the rapidly changing medium's forms and functions have multiplied and matured has defied a "proper" evolutionary cycle: "While television has begun to reach its adulthood, many forms of video are still in their infancy. Today, however, new video technologies are developing faster than theorists can examine their qualities; video may leapfrog over adolescence completely and emerge as a full-fledged, accepted medium before we ever examine what it is and what it does."[11] Like an infant entering into the symbolic discourse of medium specificity, video seems to slip among the signifiers that should resolve and name its identity once and for all:

> The change is so fast-paced that even the resources of a rich language like English/American cannot contain its demand for new terminology; the language seems to be brutalized by terms generated by the emerging and converging electronic technologies: videodisc, video-disk, vid disk, videotext, videotex, music-video, video-music, for example. We are at that clumsy stage of a toddler taking his first step.[12]

Since video's parentage appears indeterminate, its search for specificity has been equally an appeal for legitimacy. For example, critics who wish to ally video with "father film" as a medium of production, but who feel betrayed by its attachments to "mother television," may ostracize as illegitimate all consumer video formats. Sean Cubitt has aptly and succinctly noted that because its family resemblances to so many media forms prevent our tracing its filial heritage to a parent proper only to it, "video is a bastard medium, however narrowly defined."[13]

This idea of video as a bastard medium seeking its legitimate identity through its family resemblances to other media forms is a central point of departure in this and the following chapter. Here we will focus on video and television in general, and on home video and domestic TV in particular. By interrogating the residual aesthetic of negation inflecting much of video discourse, I question all custody myths that foster video's relative relation to television in absolute terms: as either rightfully separated at birth or incestuously clinging at video's expense. The primary goal is to correct binary distinctions between video and television that have marked discourses specifying the two media according to global terms. By localizing analysis within a domestic context, we may avoid the overgeneralizations and prescriptions that seem inevitable when comparing media forms too broadly—itself a problematic endeavor, since video and television, originally designed as interdependent technologies, cannot be entirely differentiated at a macro level.

My methodology will include an analysis of TV series and specials sharing the cultural functions and formal conventions of the home mode as a demonstration that, diachronically, domestic television programming since its inception has both anticipated and reconfirmed home video as a complementary medium for representing everyday life. Thus rather than jumping to conclusions that the goals of home video practice (amateur) and commercial television (industrial) are, have been, or should always be at odds (technologically, aesthetically, or ideologically), I highlight the importance of their shared domestic context to demonstrate that their development, at least on one level, has been dialogic and intertextual.

A secondary goal of this chapter is to retheorize television genres through Bakhtin's notion of "chronotope." In tracing resemblances between the family representations of home video and domestic television, we may also recognize resemblances among TV series with conventional generic identities and distinctions that would suggest greater internal differences. For example, in genres ranging from the sitcom of *Ozzie and Harriet* and the documentary of *An American Family* to the dramedy of *The Wonder Years* and the game show of *America's Funniest Home Videos,* we may observe a recollection of the past, the virtual performance of family roles, and an autobiographical desire to observe one's own mediated image. Because these common characteristics cross genres, the category of chronotope, concerned with tracing space-time relations between narrator and character and between text and spectator—rather than generic iconography, themes, or cues—is better suited as an analytical tool by which to account not only for intramedium similarities among these series but also for their intertextual relation to home mode artifacts whose conventions they share. Thus, in the broadest sense, "family resemblances" mediate the functional and formal relations among family photographs, home movies, home videos, and various domestic television programs, all of which may be grouped together as subsets of a "home mode" chronotope.

▶
_____

Video versus Television: A Residual Opposition

Since its inception, video's identity has been a point of contestation. What is its genetic background? What is its destiny? Which medium does it resemble more: film or TV? Frequently, when video displays its "maternal" identification with television, it is perceived as a regressive medium of consumption and transmission. Inversely, when it displays its "paternal" identification with cinema, video is perceived as a progressive medium of production and transformation.

For example, Roy Armes allies video with film rather than TV by privileging the substrate of videotape as the primary mark of distinction: "The new electro-magnetic recording media consist of individual works— tapes—in exactly the same way that the nineteenth-century media comprise specific photographs, films, and records."[14] Noting that television had been in full operation for more than a decade before the introduction of videotape in 1958, Armes concludes that video is in no way a prerequisite for the invention of television: "The distinction we make within the electronic production of sounds and images so as to single out 'broadcast television' is not based on the consideration of some carrier base, but in terms of social application: the diffusion of a variety of audio-visual messages from a single point (the studio) to a large number of receivers (domestic television sets)."[15] In short, Armes defines film and video as sibling media because they produce individual works (the first on celluloid, the second on electromagnetic tape) and television as a distant cousin because it merely mediates these individual works for reception through broadcast, cable, and satellite distribution—again, the opposition of production and consumption, transformation and transmission.

John Belton neatly aligns film and video as agents of production against television as an agent of distribution. When it does function as a mode of production, he argues, TV mimics cinema rather than producing its own meanings: "The technology of television has been geared to transmission, not to transformation; when it does transform, it is not television, but a copy of cinema."[16] This taxonomy grants film and video-as-film with the power to produce meaning, and television and video-as-television with the power only to reproduce meaning—to either represent or misrepresent rather than construct and negotiate social reality.

This opposition—video as producer and TV as reproducer—is fraught with unresolved theoretical problems. First, reducing video's specificity to modes of production and television's specificity to modes of distribution subjects their range of cultural effects to technological determinism. Second, this dichotomy elides fundamental comparisons that would align video and TV even at the level of technology itself: for instance, both media depend on a monitor for exhibition. Would video therefore be an immanent medium, pure only during acts of production or when stored in a series of electromagnetic signals on videotape, inevitably compromised during its act of transmission as sounds and images on television? Third, the affiliation of production with transformation and emancipation, and transmission with consumption and repression, cannot be sustained as a one-to-one correspondence.[17]

If video is a mode of production and television a mode of transmission, the crucial question arises: How may we distinguish video-as-video from video-as-television if both appear on a TV screen? Because all sorts of media texts, including videotape, telefilms, and live broadcast, appear together within the frame of a TV monitor, the only method for distinguishing a "video" artifact among them would be to know (or to imagine) the technological or institutional source of its signals. Are they "raw," direct from a camcorder (video), or "cooked" by the manipulations of post-production suites meeting broadcast standards (TV)? Are they being delivered through a closed-circuit VCR (signifier of independent or amateur status) or through commercial frequencies (signifier of professional or industrial status)? Even the general parameters of these questions are misleading, however, based as they are on the residual dichotomies of negation first proposed decades ago by video's first practitioners. Nowadays, of course, we can view commercial films on videocassette through a closed-circuit VCR or watch amateur video footage on reality-based commercial TV series. As our analysis of event video has demonstrated, formal readings alone may not be trusted to distinguish amateur and commercial modes of video production. Neither can facile dichotomies between production and consumption, transformation and transmission, or emancipation and repression be trusted to distinguish video and television. In an era of multimedia and digitization, in which film, videotape, and television have fused into a variety of hybrid forms of all ideological persuasions, conventional categories no longer contain them.

Reality-based television serves as an excellent case study by which to interrogate dichotomies pitting video and television as techno-aesthetically and ideologically disparate media. In a commercial network series, for example, such as *America's Funniest Home Videos,* that broadcasts videos produced by amateurs for entertainment and profit, how would we define the status of the individual video artifacts within each program: as home video or as broadcast television? Douglas Davis attempts to answer this question by imagining video as a transparent medium of the real and television as a manipulative medium of the fake. Although we receive both video and television on a monitor, he argues, we can distinguish video (or reality TV) whenever "the medium literally disappears" and when "the audience senses that it is beholding, not judging, a truth situated on the near side of the screen."[18] Davis indicts television's formal and commercial transformations of video as a corruption of its transparent purity: whereas video has the potential to record and transmit actuality, television manipulates video's "reality effect" for purposes of entertainment and profit:

> If we consider the precise qualities commanded by the television—and video-tape, its physical extension—we immediately see how the network esthetic has betrayed the premise, and promise, of its medium. Unlike film or print, television can connect viewers to events unfolding directly before them. Captured on videotape, which instantly plays back what it records—unlike film or print, once more—the viewer is in touch with the apparent *Real,* virtually as if he had experienced it himself.[19]

In defining video as the absence of a medium, what Davis really means is the absence of typical commercial network programming, a practice he conflates globally with television itself.

At bottom line, Davis's taxonomy yields to an unwarranted ideology of presence: inauthentic TV, like film, is prerecorded and manipulated, while authentic TV, like video, seems live and transparent. Because videotape, however, is never live itself, never less prerecorded than film, once again we see how a critic's specificity claims may be grounded more discursively in imagination than empirically in techno-aesthetic properties. Like "transparency," "reality-based" should be theorized not as a property inherent to one medium or another but as an effect of human psychology and aesthetic ideology. Yet by arguing that video is essentially more realistic than television and that commercial TV always compromises the amateur video it broadcasts, Davis mires his approach in the discourses of inherent properties and technological determinism, both of which conflate the wide-ranging fields of video and television practice to narrow structures of exhibition and sponsorship. A one-sided dialectic, Davis's oppositions move only in a single hierarchical direction: ever the cannibal, TV always cooks and consumes raw experience, always converts the video it encounters without ever being converted by video itself. By claiming that video must pursue a course entirely independent from TV, Davis theoretically divests video practices of their own power to transform televisual conventions. More important, his binary taxonomy fails to consider how the development of both media may have been mutually influential, if not ultimately inextricable.

▶───

Home Video and Domestic Television: A Dialogical Relation

The history of video need not be written solely as the ongoing negation of television. Too often, in their search to identify video's specificity, scholars and critics focus more on its disparities from TV than on its similarities. This approach must elide family resemblances between the two media, informed as it is by the modernist belief that the essence of a medium can be extrapolated from the set of characteristics present only to itself and absent in all others. But defining what is "unique" to video may not help us understand its practices—how it has been and is currently being used. On the contrary, we may become even more confused.

For example, Armes and Davis single out electronic tape as a defining constituent of video, yet they propose uses for the medium entirely at odds: Davis claims that the virtue of videotape is its capacity for the transparent transmission of reality, Armes for its formal transformation as art. The contradiction stems from their attempts to ground use value in a multivalent technological property rather than in a local field of practice, so that both construe aesthetic effects, like "reality-based," as the ontology of an apparatus rather than as the historical conventions of verisimilitude shared by a community of practitioners. The important issue should not be to define once and for all which medium is better suited to the task of transmitting or transforming "raw" reality but to trace how each medium signifies and constructs notions of "reality" itself according to adopted cultural codes. Because video and television share many of the same codes to represent everyday family life, we may learn more about use by comparing the two media than by contrasting them.

In particular, home video and domestic TV representations both share and appropriate the cultural codes of the home mode. Although studies of early television have traced the contributions of radio, cinema, and vaudeville to the medium's formal development, few have considered the legacies of family photography as an influence. During the 1950s, the formative decade of American television, snapshots and home movies had reached a peak in popularity as a form of leisure-time entertainment.

It is no coincidence, therefore, that by mid-decade, the domestic situation comedy, one of television's earliest and most influential genres, had increasingly adopted the forms and functions of the home mode as representational strategies. In series such as *The Adventures of Ozzie and Harriet, Father Knows Best,* and *The Donna Reed Show,* the American family looked back at itself, or at least the white middle-class nuclear version predominant in the postwar period.[20] Like home movies of the period, individual programs portrayed "real" families in everyday settings, sentimentalized child rearing, celebrated rites of passage, and advocated togetherness. Although radio had already introduced and institutionalized the domestic family series in the thirties and forties, television did not simply adopt their conventions wholesale. A pressing concern of the new medium would be to visualize the characters audiences had come to know and imagine only from disembodied dialogue.[21] Developed over the course of a century, the conventions of family photography provided early sitcoms with a representational strategy that could be exploited as recognizable signifiers of domestic life.

In adopting conventions of the home mode to represent its families on a TV monitor, early domestic sitcoms anticipated the advent of home video. Whereas home movies had satisfied the dream of seeing ourselves and our families on the silver screen, domestic sitcoms proxied as an expression of a similar desire for self-representation on television. If not our families, the Nelsons, Andersons, or Cleavers might provide, in the broadest sense, a vicarious experience of what it might be like to live our own everyday lives on TV, at least until the invention of consumer video technologies would allow for greater self-determination in choosing exactly how we might actually portray ourselves and our kin.[22] In this sense, we might conceive of the relation between domestic television and home video as an interplay of history, ideology, and desire, wherein each medium attempts to provide a home audience's hankering for audiovisual images of themselves, borrowing from each other over time, thus inventing and reinventing each other's conventions of representation and patterns of interpersonal communication. If television's supertext includes the family sitcom, its megatext should include family snapshots, home movies, and home video, whose intertextual relationships to domestic television programming may contribute as much to our understanding of TV's portrayal of family life as the series themselves. Before turning to a more detailed analysis of this intertextuality between domestic TV and home video, we first should establish the complementary relations between television and the home mode in more general terms.

When television was introduced to American households, it immediately evoked ideas of home and family. For example, David Morley notes that first-time owners placed family photos on top of their new TV sets to domesticate them as familiar appliances.[23] This early association of the home mode with television makes sense for a medium that specialized in serializing the private lives of its families for public display. Like snapshots, early domestic television programs shared an aesthetic and ideological commitment to intimacy, to the perception that representations of family and friends, while absent physically, seem present through their close proximity in domestic space. Although the precedence of live broadcasts at the advent of television contributed in great part to this sense of intimacy, home mode conventions helped to reinforce this impression, especially after sitcoms substituted live performance with telefilms as their standard mode of production. For example, domestic sitcoms frequently adopted qualities of home movies that draw in audience identification: direct address to the camera, quotidian activities, and real-time shooting, among others.

Like Morley, many television scholars have described the medium's domestic applications in terms recalling the cultural functions of the home mode. Horace Newcomb, for example, cites intimacy, continuity, and history as "devices that help to distinguish how television can best bring its audience into an engagement with the content of the medium."[24] Observing that TV constructs images of home that provide resources for individual identity formation, Roger Silverstone declares that the medium "marks the site of the most important *rite de passage* in our contemporary society."[25] And echoing Richard Chalfen's description of the home mode's symbolic function for self-analysis, Michael Saenz notes that "television programming serves as an ever-unfolding event inviting comparison with one's own ongoing household, and establishing a readily accessible interpretive dialectic between the two."[26]

In addition to cultural functions, television and the home mode share aesthetic qualities. Because its small screen de-emphasizes spectacle and its origin in studios set a precedent for interior settings of a domestic scale, television specialized in representing the familiar details of human behavior. John Fiske also notes that because the time taken to perform an action on TV coincides with the time taken to perform it in real life, and because the absence of editorial intervention adds the sense that the camera is merely recording what happened, television emphasizes the quotidian realism of everyday life.[27] Observing the stress on authentic details in domestic sitcoms, Nina Leibman concurs by pointing out how "studio sets boasted functioning kitchens where well-paid propmen prepared real pot roasts and mashed potatoes for fictional families."[28]

Along with verisimilitude, the domestic sitcom shares the home mode's serial aesthetic. Particularly within long-running series like *The Adventures of Ozzie and Harriet,* in which we can watch the Nelson boys grow up over the course of a decade, sitcoms present us with recurring characters that we get to know over a long period of exposure: "The constant repetition of a character means that characters 'live' in similar time scales to their audience. They have a past, a present, and a future that appear to exceed their textual existence, so that audience members are invited to relate to them in terms of familiarity and identification."[29] At the same time, the serial format's gaps occurring in between each weekly program and during commercial breaks recall the home mode's patterned eliminations of controversial or potentially offensive subject matter, such as sexuality, bodily functions, and extreme conflict. Nina Leibman points to a variety of factors impacting television's censored vision of family life, including interventions by government, sponsors, and religious organizations.[30] She suggests that the most influential technological factor reinforcing television's sense of propriety is the laugh track, "a persuasive signifier designed to dissuade the spectator from authentic feelings of grief or misery, even when the characters are experiencing such feelings."[31] In the home mode, the laugh track's corollary is the ubiquitous compulsion to say "Cheese!" thereby guaranteeing smiles for posterity rather than frowns.

During the years when Kodak sponsored *Ozzie and Harriet,* the series' credit sequences wonderfully illustrated domestic television's intertextual relation to the home mode, whose conventions were marketed to the public at the same time as they were adopted by the program as an iconographic representation of the Nelson family. After the announcer proclaimed that the "Eastman Kodak Company is happy to bring you America's favorite family," each Nelson would appear framed in the front door of their house, stopping briefly to smile for the TV camera. As they gathered together to pose in front of the shrubbery in a classic home mode group shot, the announcer would continue: "They enjoy happy times together. Like most of us, they know that good times are picture times." Suddenly the full screen image would shrink to the size of a Kodak photograph. In this transformational image, the conventions of family photography and the TV credit sequence were fully articulated and merged as an advertisement selling both the camera and the program. Introduced as having "a special idea for this weekend," Harriet in one episode appeared seated at a picnic table in the Nelson's backyard to demonstrate Kodak's new color film. Holding up photographs of each family member in typical poses, she illustrated how home audiences could model their lives and leisure activities in ways suitable for photography. By implying that the Kodak camera could preserve the American family much as television had preserved the Nelson family for generations to come, Harriet invited every family to become a TV family through the Kodacolor process, anticipating the desires fulfilled by home video several decades later.

In this example, we see how the portrayal of the Nelsons as both a genuine and a TV family was modeled not only on their actual home life but on the conventions of family albums of the period. Because they had performed for years on radio as disembodied voices before their television debut, the home mode offered a popular, recognizable set of representational codes with which to visualize the relation of each family member to his or her domestic space. In particular, home mode conventions reinforced an image of the Nelsons as a "real" family living in a "real" suburban home.

Like many early television performers, such as George Burns and Gracie Allen or Lucille Ball and Desi Arnaz, Ozzie and Harriet Nelson played versions of themselves, even casting their sons David and Ricky to play their children. Blurring fiction and nonfiction, in virtually every credit sequence, the camera glorified the Nelson homestead, supposedly modeled on their house in Hollywood, as the stage on which the family could perform itself. An announcer would then introduce each Nelson as a composite of person, actor, and character (e.g., "Here's Ozzie, who plays the part of Ozzie Nelson"). Shot in medium close-up, pantomiming dialogue and smiling for the camera, Ozzie and his clan would impersonate their real-life identities before acting them out as if in a televised home movie.

By analyzing credit sequences as pithy statements of television's evolving idea of what family life in the 1950s might look like for broadcast, we see how the home mode may have served as a primary influence on televisual codes of representation. We also recognize that "the Nelsons" were as much a discursive construction as a real entity: the resemblances between Ozzie, Harriet, David, and Ricky as historical personages and as fictional characters deconstruct conceptions of "the family" as some ontological norm existing outside the process of signification against which its representations should be measured. In addition, the mediated resemblances between the Nelsons and the American families who may have emulated them in their own snapshots and home movies suggest that the social reality of everyday living consists not in some empirical actuality that television and the home mode either reflect or distort as secondary or derivative representations but in the dialogue between their systems of signification from which the idea of the family is imagined and generated.

Professing that pictorial communication structures our dialogue with the world, Sol Worth writes:

> Correspondence, if it makes any sense as a concept, is not correspondence to "reality," but rather correspondence to conventions, rules, forms, and structures for structuring the world around us. What we use as a standard for correspondence is our knowledge of *how people make pictures*—pictorial structures—how they made them in the past, how they make them now, and how they will make them for various purposes in various contexts.[32]

Let us turn, then, to some of the conventional correspondences mediating and structuring the resemblances between the families of domestic television and the home mode.

▶
───────────────────────────────

Family Resemblances: Mediating Correspondences

In her cultural history of early American television, Lynn Spigel argues that the medium's ideological relationship to "the family" was dialogic rather than deterministic. Adopting an approach rejecting essential oppositions between televisual images of the family and the "real" family itself, or models positing a direct cause-and-effect relation between them, she analyzes television's discursive strategies that signified and dramatized domestic arrangements contributing to the audience's image of family life. Focusing on nascent genres to locate the origins of these strategies, she finds correspondences between TV and social reality guided by conventions joining home viewers to the fictional homes being broadcast, in particular the aesthetic codes of immediacy, presence, and intimacy that inscribed viewers into the programming's narratives, thereby bridging their imaginary distance.

Rather than as a window onto reality, a common metaphor in the 1950s, Spigel reconceives the television screen as something akin to an interactive proscenium inviting viewers to imagine themselves "*on the scene* of presentation."[33] She identifies two seemingly contradictory impulses such a scene evokes in typical early sitcoms: a theatrical sense of the situation's exaggerated performativity and a parasocial belief that its characters are real.[34] Spigel's dialogical approach is significant in its emphasis on the intersubjective relation between conventions of media representation and spectatorial practices. Neither entirely outside the televisual text nor reflected in it, the family's image of itself fluctuates in an intertextual context that, not limited only to TV, includes other popular discourses such as interior design magazines, lifestyle advertisements, and homemaking advice columns.

Among these discourses is the home mode, briefly mentioned but largely overlooked by Spigel. As a coded system of domestic signification common to both empirical and television families, the home mode bridges the desire of the former to appear like the latter by visualizing domestic living as a scene of presentation that family members may perform on (and in), as well as view from a mediated distance. Discussing television's parasociality, M. S. Piccirillo compares its viewer's inclinations toward observation and participation in an explicit analogy to home mode spectatorship:

> Viewers are positioned to *observe* the difference in the episode by discourse which presumes that they are experiencing the end of a regularly scheduled viewing habit. At the same time, viewers are positioned to *participate* in the similarity of the episode by discourse which presumes a close knowledge of the history of the series. The experience is analogous to perusing a family photograph album: One observes clearly re-presentational photographs and participates in the memories those icons revive.[35]

Although mediated, like domestic sitcoms, home mode artifacts rely on ideologies of presence and intimacy. Psychologically, their spectatorial relations simulate face-to-face interaction by acknowledging that their subjects of representation exist and can be located physically outside of signification in the social worlds of their invested viewers. Therefore, by adopting the home mode's conventions to prop up a perception of immediacy, fictional TV families could replicate its existential logic. This impression that television's characters make up real families who happen to live out their lives on TV encouraged viewers to imagine the Nelsons, for instance, as next-door neighbors, even though their scene of presentation was actually confined to a Hollywood studio.

At the same time that the domestic sitcom's conventions of parasociality could naturalize studio sets as real homes, its conventions of theatricality could denaturalize real homes as prosceniums staging everyday perfor-mances of familialism. Noting that many programs "self-consciously ac-knowledged the theatrical artifice involved in representing a naturalistic picture of domesticity," Spigel suggests that the sitcom provides a potent metaphor for the spectacle and social performance typical of modern fami-ly living, whose self-reflexive nature often requires that members act out fabricated roles.[36]

Echoing home mode conventions for staging family relationships, sit-coms use architectural elements of the household, such as windows, door-ways, and foyers, to position the actors in proscenium space. For example, in the credit sequence from *Ozzie and Harriet,* the front door works like a bordered frame to represent each Nelson as an individual, and the front lawn and shrubbery provide a backdrop to represent them together as a group. This mannered form of displaying the family recalls the artificial poses typical of family photos and provides, according to Spigel, some criti-cal distance from everyday life.[37]

Therefore although both domestic television and the home mode may idealize family life, the performativity required of its participants for re-cording its portrayals may denature any sense of their so-called ideological effectivity. Countering arguments that home video, for example, uncon-sciously reproduces the nuclear family as a given, we may demonstrate its capacity to trigger self-awareness of the arbitrary nature of domestic role-playing. Psychologist L. S. Vygotsky has observed, in fact, that being and playing family members coincide in the development of children's capaci-ties to recognize behavioral norms as discursive constructs. Discussing the importance of playacting to cognitive development, he refers to a study in which two biological sisters, aged five and seven, "played sisters," thus making their game resemble reality. The vital difference, he claims, is that in self-consciously displaying their sisterhood, the girls attempted to re-enact what a sister should be. As a result, each came "to understand that sisters possess a different relationship to each other than to other people. What passes unnoticed by the child in real life becomes a rule of behavior in play."[38]

By extension, playing sisters on home video may reinforce the child's self-awareness, both during the process of recording, which requires she don a sisterly persona, and the act of reception, which provides the critical distance for her to reflect on and evaluate the success of her performance. Presumably, therefore, the home mode may either reinforce or revise her behavior, depending on her positive or negative reaction to the video's ex-ternal correspondence to her internal self-conception. Like home video, do-mestic television may also function as a measure of each family member's ego ideal. As a series entitled *Sisters* illustrates, actors playing sisters may serve for viewers as models of sisterly behavior through processes of spectatorial identification.

Family resemblances, it must reiterated, refer to fuzzy relations, not mirror images. Because the symbolic codes and media conventions that make up television's representations of the family cannot be equated with the material culture of everyday life, a program's "construction of the familiar life-world of the ideal viewer will be seen as always being at some distance from the real viewer's interest-related interpretation of the rules, roles, and regularities of the life-worlds in which he or she actually exists."[39] When analyzing family representations, therefore, either on TV or in home videos, we must always consider the historical moment and cultural context in which these rules and roles hold precedence, their formal and ideological fluctuations over time, the popular and organic memories of family preceding them, the intertextual representations of family produced by other media forms, as well as the contemporaneous social configurations of actual families themselves. For example, in evaluating portrayals of the Louds in *An American Family,* we would be naive to measure them simply against statistical norms or some timeless, essentialized notion of "the family." In addition, we would have to take into account the changing cultural history of American families; our memories of televised representations of family life preceding the documentary, such as *Ozzie and Harriet;* comparisons to popular television programs, such as *All in the Family* or *The Brady Bunch,* as cultural touchstones of the same period; representations in competing media forms, whether home movies, cinema, literature, or advertising; and, finally, less tangible determinants, such as the legacies of oral history, organic memories of the past, and personal feelings about the present. Likewise, in evaluating a series such as *The Cosby Show,* another landmark domestic sitcom, we would be somewhat amiss to measure the portrayal of the Huxtables strictly against the statistical norms of contemporaneous African American families, without considering the genealogy of popular representations of family with which that series negotiated.[40]

In short, questions of resemblance are subject to diachronic transformation. Although family representations may change in direct response to actual social alterations in domestic living arrangements, as often they deflect or resist such changes by reiterating residual conventions behind the times. Thus, for example, although by the mid-1980s the nuclear family model no longer held sway as an American hegemonic norm, a resurgence of television series, from *Family Ties* to *The Cosby Show,* featured domestic portrayals reminiscent of the 1950s. This "lag" should suggest not that these series were "inaccurate" or "inauthentic," terms irrelevant to symbolic world making, but that images and social prototypes of the family are better kept distinct precisely because of their historical disparities, which should prevent holding one prototype accountable to another according to the terms of reflection theory. In the same way that a proper history of the family must avoid teleology and functionalism to explain its evolving relations to social and cultural formations, so must a history of family representation.

For the remainder of this chapter, I will demonstrate the intertextual and diachronic processes of signification that have constructed and contributed to the social reality of "the American family" over the past four decades by tracing a selective history of landmark television series sharing acute family resemblances: *The Adventures of Ozzie and Harriet, An American Family, The Wonder Years, The Real World,* and *America's Funniest Home Videos.* The goals of this history are threefold: (1) to foreground the influence of home mode conventions on domestic television's codes of representation; (2) to trace the development of this influence historically by noting the resemblances and differences among the series themselves, as well as their relations to the evolving practices of the home mode and to actual social changes in family arrangements; and (3) to classify these series not according to traditional TV genres but according to Bakhtin's concept of chronotope.

▶

The Adventures of Ozzie and Harriet: Sitcom in the Home Mode

A paradigmatic situation comedy, *The Adventures of Ozzie and Harriet* became the first television series to experiment significantly with the visual conventions and cultural functions of the home mode to portray the iconography of the "typical" American family.[41] Already famous for their radio broadcasts beginning in 1948, the Nelsons were known in 1952, the year of their TV debut, as musical entertainers. In the transition from radio to television, Ozzie and Harriet appropriated home mode conventions purposefully to reinvent their glamorous image as celebrities into a more ordinary image as middle-aged spouses and parents whose daily life at home raising the kids just happened to be televised for public viewing. Dropping almost all references to their show business careers and locating their homestead in a typical Los Angeles suburb, they emphasized light humor arising from everyday problems and simple misunderstandings among family, friends, and neighbors.

Although consistently the most watched program in its time slot and one of the longest-running series in television history, *Ozzie and Harriet* irritated many critics unsure of what to make of its quotidian premise, which in comparison to the slapstick humor of variety shows or the intense drama of live TV seemed mundane, bland, and boring. For example, Rick Mitz has belittled the series "because it took family life—a hotspot of American insecurity—and trivialized it to the point of innocuousness, while squeezing out all the humor like puerile pulp. *Ozzie and Harriet* was the ultimate in home movies—with an annoying laugh track."[42]

Of course, like criticisms of the home mode discussed in chapters 2 and 3, Mitz's barb betrays his own prejudices against home movies by holding their similarity to *Ozzie and Harriet* as somehow accountable to what television should more rightfully be. He also misconstrues the intentions of the series, more innovative at the time than was apparent. Unlike TV's comedy teams such as Burns and Allen, or Lucy and Desi, who evoked their status as real-life married couples primarily to suspend disbelief in their absurd situations, Ozzie and Harriet attempted to make their real-life and TV identities coincide so that the plausibility of their family would appear more important than their "adventures." Ozzie himself has justified accusations of blandness by upholding the adherence of his scripts to the reality of autobiographical experience: "Being a family in real life, our relationships are honest and the viewers are able to believe what they see and associate our problems and activities with their own lives."[43] Critics more sympathetic than Mitz concurred. One review in *Variety* praised the series because "what sets the Nelsons' show apart from others of its ilk is that it's pure family, not fiction."[44] In *TV Guide,* Cleveland Amory wrote that "it is not only a wholesome show for the whole family, it *is* the whole family."[45] By the early sixties, so indelible were the parasocial relations establishing their verisimilitude that the Nelsons were introduced at the beginning of each episode as "America's favorite family."[46]

The primary significance of *Ozzie and Harriet* is that more than any sitcom of its time or after, it emulated the home mode's referential and cultural functions of autobiography, generational continuity, group cohesion, and nostalgia for the past. One episode, entitled "Little Handprints in the Sidewalk," summarizes the main concern of the series with memory and its indexical connection to family representation. The plot, typically, is quite simple: while replacing their cracked sidewalk, Ozzie and Harriet debate the merits of saving a slab of concrete preserving imprints of the boys' hands taken when they were very young, for the handprints serve as lasting reminders of past memories. In one exchange, Ozzie directs Harriet to "look at the size of those hands," to which she replies, "I'd almost forgotten they were ever that small." After the entire family decides to keep the slab, they work together to find a place for the bulky memento, but no room in the house seems able to accommodate its size and weight. Agreeing to no other option, finally, than to pulverize the concrete with his sledge-hammer, Ozzie is caught short by a hallucinatory image of his sons' faces superimposed within the handprints, looking up at him as if appealing for a stay of execution. Unable to shatter both his memories and their imprinted signifiers, Ozzie donates the slab to a neighbor for use as an anchor.

An allegory for indexical processes of human representation, the handprints recall André Bazin's diagnosis that the plastic arts are motivated by the drive for preservation. Ozzie links the imprints so strongly to the actual physical existence of David and Ricky that he cannot destroy them. On the other hand, while cherishing the handprints, Harriet understands that their time has passed: "It's a shame to lose them," she remarks, "But I'm afraid that they have to go. Time marches on. Make way for progress." In the Nelson household, this progress in family representation is marked conspicuously by ubiquitous snapshots and home movies. If a concrete slab no longer fits into the efficient designs of their suburban domestic space, how simpler it is for the Nelsons to inscribe their memories in the stream-lined pages of a family album, whose indexical images are not only easier to store and access but iconographic to boot.

Like Bazin in spirit, Ozzie and Harriet consistently celebrated photography as the supreme medium of preservation. Throughout their fourteen-year run, the Nelsons would frequently review or refer to their personal home mode collection, in some episodes ignoring the conventions of plot altogether merely to take time out to reminisce. For example, "The Big Dog" begins with Ozzie and Harriet addressing home viewers from the comfort of a couch: "We've just been looking through some of our old photograph albums. They bring back some wonderful memories, but they also remind us of how quickly time passes." Identifying David on his second birthday, they remember traveling to Columbus, Ohio. Turning to pictures of Ricky on his first birthday, they try to recall his dog's name. Husband and wife continue for some time in this casual, meandering fashion. These recollections only vaguely connect to the episode's larger narrative purpose; instead, they simply reiterate the pleasures of looking at family photos.

Not just a sitcom *in* the home mode, The *Adventures of Ozzie and Harriet* is *about* the home mode as well. It was no coincidence, therefore, that Kodak sponsored the series for several seasons. In many of Kodak's commercials, the Nelsons would demonstrate the manufacturer's new equipment by photographing each other at home, then review the results by leafing through snapshots or gathering to watch home movies in the den. As a real family recorded on TV recording each other on film both in their adventures and in commercials, the Nelsons blurred the typical distinctions between social reality, nonfictional advertising discourse, and fictional narrative, all merged seamlessly within a unified televisual mise-en-scène. Ozzie, Harriet, and "the boys" sold Brownie cameras and film from their backyard, kitchen, and automobile, spaces in which the pursuit of family life as their primary form of recreation seemed unmediated in both ad and story. Where the real Nelsons, their characters, and their corporate personae began and ended could never precisely be drawn.

For example, an ad for the Brownie Starmite at the beginning of "The Girl in the Emporium" illustrates the embedded intertextuality that the series depended on to portray the illusion that the Nelsons' life and art were one and the same. The spot begins with each family member stepping out the front door of their home, Ozzie carrying a camera, David a Christmas wreath, and Ricky a hammer and nail. While the boys hang the wreath, Ozzie takes their picture as Harriet looks on. This typical holiday scenario rapidly becomes transposed into a framed photograph. We then cut inside the house to Harriet, who turns from wrapping packages toward the audience to recommend a new Kodak camera as a perfect gift: "Imagine that fun Christmas morning when someone finds a Brownie Starmite outfit under the tree." Here Harriet appears to be playing spokeswoman rather than wife and mother, but the subsequent cut to David, who opens a present on Christmas Day to reveal the camera Harriet had just been selling, merges family and corporate identities, private and public display, gift and commercial exchange. As David begins to photograph the rest of the Nelsons around the tree with his new Starmite, Harriet's voice-over continues: "You can picture the fun of Christmas right from the very first moment, and from Christmas Day all through the year this will be a gift the whole family will enjoy and appreciate." And in fact, soon enough we see the Nelsons leafing through snapshots of Ozzie holding golf clubs, Harriet gardening, and the boys vacationing in Hawaii. Coming full circle, bridging past, present, and future, the ad erases spatial and temporal boundaries that identify it merely as an ad, rather than as a typical scene we might catch the Nelsons performing at home or on their program.

Whereas this strategy may have naturalized Kodak's commercial discourse in hopes of increasing retail revenues, more interesting is how it taps into the desires of home audiences to emulate the Nelsons as a family at once ordinary and special. Identifying with common behavior such as gift giving and holiday cheer, audiences may have felt empowered to imagine themselves, too, someday on television, particularly if the Nelsons seemed extraordinary only in that they had managed somehow to realize this dream already. For the time being, at least, home movies could substitute as the next best thing. If not in their own TV series, like the Nelsons, viewers could star in movies of their own making. Kodak's commercials emphasized this connection, noting in one spot how "the Nelsons always bring along a movie camera when there's fun in store because they've found that big moments just can't get away when you save them in action and color. . . . Right now is the time to start making your family movie stars."

By explicitly linking the performative elements of home movies to those of the domestic sitcom, the Nelsons foregrounded the theatricality of family living. Inviting home viewers to follow suit by exploiting their own families as a mediated form of entertainment, Ozzie himself served as the perfect model for fathers and husbands who, during this period, generally took command of home mode practices. Like most home moviemakers, Ozzie reserved for himself complete creative control, in this case of a television series: directing and producing nearly every episode, approving all script outlines before their completion, and exerting final authority on the soundstage and in the editing room. In fact, he obtained the first non-cancelable ten-year contract in the history of show business, signing with ABC in August 1949.

A consummate businessman, Ozzie nevertheless played down his professional status to reinforce viewer identification with his "average Joe" persona, an amateur at heart. In several Kodak commercials for home movie cameras, Ozzie would appeal to his authority as a filmmaker to sell their "expert" settings at the same time he would admit to his personal preference for the ease of their automatic controls. For example, in one spot for the Brownie Turret broadcast during an episode entitled "The Bridge Group," he demonstrates the camera's close-up, medium, and wide-angle lenses by shooting David and Ricky playing badminton in the back-yard. After authenticating the camera's value with a personal endorsement ("Here's the Brownie Turret I use"), he then domesticates the technology for less proficient technicians at home: "I didn't have to move a step. What could be easier?" Indeed, like Ozzie, the viewer too can make home movies "with a real professional twist," without leaving the house or backyard.

A television series that asked home audiences to conceive of its representations as extensions of their own home mode practices, *Ozzie and Harriet* augured the popular desire of home video viewers finally to watch their own families directly on TV rather than a home movie screen. This longing, still unrealizable during the fifties and sixties when the series aired, was frequently illustrated within Kodak ads by editing devices conflating home movie screens and snapshot borders with the frame of the television monitor, or special effects that animated photographs without the use of motion picture technology. For example, in one spot broadcast during an episode entitled "Rick's Scientific Date," snapshots come to life like a home video. While a plaintive chorus sings "Wish you were here," an elderly woman pauses in her parlor to admire two large framed photographs of her grandchildren. Soon the mailman rings her doorbell, delivering a letter filled with new family snapshots, described by the spokesman as "the next best thing to a personal visit." As the grandmother props one of the snapshots up against the base of a table lamp, the spokesman continues: "To one who's waited so eagerly to see them, snapshots will almost seem to come alive." Suddenly the static image of two toddlers at play moves like a miniature home movie, but without requiring darkness, a projector, or a special screen. While activated by a desire for the living presence of their subjects, these "magic" photographs may also signify a wish for a new form of animated home mode artifact as easy to view as, say, popping a tape in the VCR. Wouldn't it be wonderful, the ad suggests, if we could observe moving images of our families, no matter how distant, with the same ease as turning on TV to watch the Nelsons?

Ozzie and Harriet, in fact, often felt the same way about their own series, treating previous episodes as we treat our home video libraries today, rerunning "flashbacks" to relive memories of earlier times. For example, during an episode entitled "The Little Visitor," in which neighbors leave their young son with the Nelsons to baby-sit, Harriet introduces David and Ricky from a sequence that had been broadcast years earlier, inspired more by her nostalgia than by the episode's narrative logic. Addressing the audience directly, Harriet says wistfully: "You know, when Ozzie and I were talking about David and Ricky having their slight altercations when they were little, we weren't kidding. Here's a sample of what used to happen about ten years ago." Taken out of its own narrative context, the clip that follows relinquishes its function as fiction, transformed into a pure home mode moment recording how the boys have grown and matured over the years. As Harriet reminds us, "It hardly seems possible they were really that small once, does it? We love those old pictures. We hope you do too." Here she could not be any more explicit about a primary function of the series as a documentary of her family, whose general course of development seems more important than their individual adventures.

In an episode entitled "David's 17th Birthday," the sitcom's function as family archive reaches its most sophisticated level of intertextuality. It begins with a Kodak commercial for the Brownie Turret. While visiting Disneyland with the boys, Harriet demonstrates the camera: "We're having a wonderful time. And best of all, we'll be saving all of it to enjoy again and again." The scene then cuts to the Nelsons at home watching these same movies in their living room. Like an epigraph for the story that follows, the ad succinctly enacts a metadrama epitomizing the fundamental intention of the series toward mediated self-reflexivity.

Ozzie then appears as a master of ceremonies, announcing that the episode, airing in 1962, was actually recorded in 1955: "In case we all look a little younger on tonight's show, it's because it was filmed nearly seven years ago. In fact, it's all about David's 17th birthday." As a filmed rite of passage reviewed on television to remember the past, the episode is comparable to a film-to-tape transfer of home movies. Like a typical home movie narrator concerned with people more than their situations, Ozzie provides running commentary on the action in voice-over, pointing out that "this smiling young fellow is David as he was in those days," and inquiring if we recognize Ricky "when he looked like this."

Later in the episode, after the family celebrates David's birthday, filmed with typical home mode conventions such as singing "Happy Birthday," presenting a cake, and blowing out candles, the Nelsons retire to the den to watch old home movies of David as an infant and toddler. This sequence is extraordinary for early television, as the images are authentic, and thus genuinely moving to the Nelsons and to the home viewers who have come to know the family intimately over the years. In particular, the movies raise curiosity about what the family looked like before their appearance on TV, or even radio itself. Observing that "these films have sure held up well," Ozzie notes that one scene is from 1938, exclaiming, "Hey, that's me!" as his dapper image as a new father appears on the screen. This response, "That's me," tersely expresses the act of self-recognition underlying the sitcom's relation to the home mode, here represented by Ozzie at different stages in his life cycle from the late thirties to the early sixties, when he returns at the end of the episode to bring the action back into the present.

While poignant, the sequence is also humorous. Taking over Ozzie's earlier function as narrator, Ricky adds his own comic twists on the events projected. For example, poking fun at David as a fat, hairless infant, Ricky jokes, "Hi ya, baldy!" and "Weren't you a chubby little rascal?" As he later explains to his mortified brother, "I was only trying to liven things up." Ricky's acknowledgment that home movies can often be better appreciated as entertainment by adding comedy or foregrounding the subject's embarrassment has been taken to its logical conclusion three decades later by Bob Saget, who as host of *America's Funniest Home Videos* interspersed quips between clips over which he voiced brief parodic story lines. Like Ricky, Bob had become part of the nation's "TV family" over the course of eight seasons, not to mention five more as a father on *Full House*. His barbs therefore seemed like familiar ribbing rather than estranged cruelty, and what's more, if we were lucky, it might have been our very own family and friends that he was cutting up for our amusement.

As noted earlier, though known for rather mild humor, *Ozzie and Harriet* did at times feature slapstick sequences also reminiscent of *America's Funniest Home Videos*. For example, the end credits of several episodes rebroadcast particularly silly moments for another laugh: embarrassing accidents, such as a dish falling on Ozzie's head; sports bloopers, in which the boys knock Ozzie about while playing football; and unexpected outtakes, such as a fire flaring up in a prop wastebasket. Ozzie would also experiment with special effects now ubiquitous to VCR controls: a slow-motion rendition of himself falling headlong into a cake; a reverse-motion sequence of David jumping from a tree; or a forward-reverse loop of himself drinking a glass of water. If many sequences in the series pointed back to the residual pleasures of home movies, these interludes pointed ahead to the emerging innovations of home video.

In both content and form, *The Adventures of Ozzie and Harriet* explored mediated performances of familialism with more determination than any other series with which it shared the airwaves. Yet despite its self-reflexivity, like the family albums and home movies typical of its time and place—the postwar American middle-class suburbs—the series literally "whitewashed" everyday life. In displaying themselves, the Nelsons portrayed the family they would like to be (or to be imagined as), thus editing out dysfunction and unhappiness to preserve a domestic ideal. Through its own set of patterned eliminations, such as the repression of all references to Ozzie's current line of work or the bracketing of action to weekend leisure pursuits, the series rarely ventured for long beyond the safe haven of the Nelsons' front yard.

> The family-centered television situation comedies of the 1950s were similar in the way they *seemed* to escape the intrusion of the state. The state, civil society, even the common world of work were alien activities existing somewhere "out there," beyond the pleasantly manicured suburban lawns. Politics, by its telling absence, was a contaminating force to be kept beyond the threshold of the private home.[47]

As a prototypical ideal, then, the Nelsons have accumulated immense iconographic value. Their series continues to stand in the popular imaginary for an idea of "the American family" against which subsequent families have been measured. Its home mode aesthetic may now appear naive, nostalgic, or sentimental, but as an ongoing documentary of a single family, *The Adventures of Ozzie and Harriet* set an important precedent for television to emulate, revise, and repudiate in the decades to follow.

An American Family: Documentary in the Home Mode

Produced in 1971 by Craig Gilbert for American Educational Television, *An American Family* aired on PBS for twelve weekly hour-long episodes documenting the domestic lifestyle of the Loud family of Santa Barbara, California. Writing in *TV Guide,* Margaret Meade championed the series as an anthropological field study, a "new art form" in which the Louds had agreed "to lay their as yet unlived lives on the line, to share joys and sorrows they had not yet experienced."[48] Meade's enthusiasm may be explained, on one hand, by her approval of the increasingly widespread adoption of cinema verité as a methodological tool of ethnographic study and, on the other, by her surprise at a depiction of the American family without the patterned eliminations of controversial subject matter typical of television. *Time* magazine framed the same issues more in terms of pop culture, noting that although the Louds may look like the cast of a family sitcom, their story was "no Ozzie and Harriet confection."[49]

Although the *Time* story cited *Ozzie and Harriet* to make a distinction between the two series, it would seem that the Nelsons share a great deal in common with the Louds: two real, white, affluent nuclear families situated in Southern California, whose everyday, frequently mundane lives generally play out in their suburban home on television. The difference, and crux of Gilbert's argument, is that the Louds' efforts to uphold and preserve the family ideal represented by the Nelsons have met increasing challenges from changing social mores generated by the cultural upheavals, liberation movements, and economic crises of the late sixties and early seventies—conflicts of familial interest that Ozzie and Harriet, already canceled by that time, may never have imagined. The sociopolitical world that the Nelsons had managed to keep outside the borders of their white picket fence has invaded the domestic haven of the Louds, internally by problems such as divorce, homosexuality, and labor disputes, and externally by the documentary crew itself, prepared to make public their very personal affairs. If Ozzie also shared the Nelsons' private moments with the public, he at least controlled the form and content of his family's image.

Rather than think of sitcom and documentary as fundamentally different, it makes more sense to view *An American Family* as a continuing investigation of the hegemonic family paradigm established and mediated by *The Adventures of Ozzie and Harriet*. Pat and Bill Loud suffer not merely because they must deal with adultery and apathy but because their family trajectory increasingly diverges from the nuclear ideal of family togetherness sentimentalized by the domestic sitcoms and home movies that they grew up imagining as the straight and narrow path to domestic bliss. Therefore the central tension of the series arises from the conflict between the social reality of the Louds and their predominant cultural image of family.

Dan Graham cites the same gap to explain adverse critical reactions to the series:

> An idealization of the American family was placed in doubt: The Louds were thought not to be typically representative of the American family. The program or the Louds' life-style were termed "negative"; the editing was considered biased. Initially, critics thought the show to be an attack on the "shallow, petty and materialistic" aspects of the American family or of only the affluent California family. They saw the Louds as "all too typical but not like us" *or* totally atypical and a deliberate misrepresentation of the normal American family.[50]

Suggesting that our conceptions of family are indelibly marked by popular memories of its mediated representation, Graham further observes that "as the series progresses we see more and more the differences between the families of TV and the families of reality."[51]

In fact, Alan and Susan Raymond, who spent seven months shooting footage of the Louds, self-consciously conceived the documentary as a revisionary take on the domestic sitcom:

> The television series about a family has been a classic staple of TV almost since its inception. We all remember "Father Knows Best" and "Ozzie and Harriet." But in those kind of shows the family was depicted in a very idealistic manner. Sure they had their problems, but they always got solved in a very pat way. And though we might all like to think of our fathers or mothers as Robert Young or Harriet, and our younger brothers as Ricky, none of us really had those kind of families. What happened is that an entire generation of viewers was unconsciously traumatized because they could never measure up to the image of family life they saw on the screen.[52]

Gilbert's underlying thesis as the Raymonds' producer was to deconstruct the myth of "the" American family: "If I film any one American family over a long period of time, I will expose the myths, the value systems, the ways of interacting that are American and apply in some ways to all of us."[53] Emphasizing the Louds as merely "a" family, he exploits their similarities to the nuclear ideal in order to shatter it. By juxtaposing conventional representations evocative of domestic sitcoms and home movies with documentation of the unpleasant scenes and behaviors such artifacts typically edit out, Gilbert subjects the Louds to ironic scrutiny without having to add voice-over commentary. Moving between consensual and critical perspectives, Gilbert fashions a documentary practice informed by and deforming the home mode. Pat Loud, when first hearing about the project, made the connection explicit: "We went into this thinking super home movies, neat!"[54]

The aesthetic of Gilbert's documentary practice shares many intentions of the home mode's realist folk myth, best defined by Richard Chalfen:

> The photographer is conceived of as witness, and photography is primarily reportage; photography is grounded in theories of empirical truth, informative value, and finally metonymic significance. In this realist perspective, photographic images come very close to copying reality and "standing for" the things, people, place that appear in pictures.[55]

This description applies equally to Gilbert's "observational mode," which stresses the absent presence of the filmmaker and editing that enhances the impression of real time, immediacy, and intimacy.[56] Here the parasociality providing the Louds verisimilitude stems not from techniques such as address to the camera, typical of *Ozzie and Harriet,* but simply from the fact that the Louds have invited the viewers into their real home. Although the Nelsons may have been a real family, they played themselves in a replica of their house on a Hollywood studio set. Where we could only imagine them as next-door neighbors, we might actually visit the Louds in Santa Barbara. This life world connection between the Louds and the American families at home watching them reinforced viewer identification. According to the Raymonds (echoing Ozzie Nelson), "The audience sees real people, not actors, and hopefully through watching the Louds sharing familiar experiences the viewer can learn something about himself and his own family."[57] Gilbert concurred: "A surprising number of viewers around the country have felt an identification with the Louds. Thus the series helps people see themselves and their lives from a fresh and slightly different point of view."[58]

Discourses such as the Raymonds' that the Louds are "real people, not actors," or Graham's that they represent "families of reality," not "families of TV," attempt to erase the documentary's formal mediations in favor of its supposedly realist aesthetic. But aren't the Louds, after all, a television family supreme? Don't we come to know them *only* through the medium? Even the Nelsons had a public existence outside of TV before the debut of their sitcom. With no external standard by which to measure the authenticity of the Louds' on-screen behavior, we should be hard pressed to judge when they are being themselves (real people) or playing themselves (actors). Although the series calls upon the transparent aesthetic of documentary's observational mode, the presence of the camera introduces an impulse toward theatricality, toward performing an idea of the self, an idea of the family. Richard Chalfen, although generally regarding home mode artifacts as sociological data, acknowledges that "picturetaking has the power to transform on-going patterns of activity into other behavioral routines—into patterns of behavior that are socially appropriate and culturally expected when cameras are in use."[59]

If *The Adventures of Ozzie and Harriet* illustrated the proclivity of home mode practice and the domestic sitcom toward self-conscious theatricality, *An American Family* illustrates the same capacity in cinema verité. That the Louds rarely acknowledge the camera cannot be construed to mean that the filmmakers' presence did not inform the family's performance as subjects. Pat Loud admitted:

> Cameras manipulate people. I don't mean to say that ever and always we were fully aware that the camera was on us, but the brain never fully rejects that kind of information. Cinéma vérité, maybe; sound sociology it isn't. . . . The cameras, as I say, made a difference in how we acted toward one another; you might imagine its effect on a bunch of teenagers brought up on television.[60]

Understanding the influence of televised images, Pat suggests that her children may in part have been performing themselves according to an imaginary script penned by intertextual references to other mediated family representations. Like the Nelsons, the Louds on camera may resemble themselves off camera, but are they truly self-identical? Bill Nichols refers to this paradoxical nature of documentary subjectivity as "virtual performance," which "presents the logic of actual performance without signs of conscious awareness that this presentation is an act."[61] Perhaps a fundamental difference between the Nelsons and the Louds is that Ozzie foregrounded and celebrated the suburban family's virtual performances, whereas Gilbert disguised them, despite acknowledging later that the Louds "all became quite sophisticated about the camera and realized that every ten minutes the film would have to be changed. During the changes they would talk about the things they considered too private to be filmed."[62]

Here we see how the home mode's patterned eliminations did in fact exert influence on the editing process. Indeed, the perspective from which critics viewed the structuring absences of the series strongly influenced their evaluation of its cultural function. Douglas Davis, for example, who championed the series for its critical function as documentary, felt that "though edited, it made less attempt to structure and pace narrative events than any popular television series yet. Often long stretches of meaningless, boring conversation were allowed to play out, unstructured."[63] Pat Loud, however, who volunteered to participate in the series for its consensual function as a home movie, felt betrayed when "all of the joyous, happy hours of communication and fun were left out."[64] She did note, however, that the Raymonds and Gilbert "were very good about cutting some things that we said would have been unpleasant or would hurt somebody's feelings."[65]

In a sense, then, because they were welcomed by the Louds over the course of the production as surrogate kinfolk, the Raymonds functioned as much like home moviemakers as they did documentarists: "The Louds . . . accepted us as members of their family. None of this would have been possible if we hadn't formed an intimate relationship with each member of the family, a bond of mutual trust and respect. This enabled us to film extremely sensitive moments in their lives."[66] Arguably, this bond of trust might also have tempered their documentary's critical distance by whitewashing its "raw" footage with patterned eliminations of "objectionable" content, further aligning the series with the home mode and domestic television programming.

This oscillation between consensus and critique is the series' structuring dialectic, moving between both poles over the course of twelve episodes. Particularly in the first third of the series, the episodes recall *Ozzie and Harriet* in form and, to a lesser degree, content. The first episode portrays, for instance, as in so many sitcoms, wife and mother preparing breakfast for husband and children before they leave for the public world of work and school. Like the Nelsons, Bill is the breadwinner, and Pat the housewife; the attractive kids are presented as having average intelligence and typical extracurricular activities. Although the Nelson family lacked girls, Delilah and Michelle live, according to brother Lance, a "Tammy existence" evocative of the 1950s. And like Ricky Nelson, Grant aspires to a career in rock and roll and performs at the end of the episode. Similarly, in the paradigmatic style of *Ozzie and Harriet,* the fourth episode begins with an establishing shot of the Loud home followed by the signpost revealing their family name, which compares the series to its domestic sitcom forebears while setting it up for the contrast to follow.

The producers ask us to think of the Louds through the conventional ideas of family we have garnered through its idealized representations that they will ultimately shatter. The recurring credit sequence suggests this movement from unity to fragmentation. A variation on the opening of *The Brady Bunch,* an idealized domestic sitcom popular at the time of the documentary's broadcast, it inverts the premise of that series about two fractured families (caused by the death of a spouse) who unite through marriage, each family member representing a subplot (illustrated by individual boxes) in an overarching grid that joins them as a single entity, as an incorporated American family. By echoing this credit sequence, but adding the splintered graphic "FAMILY" at the end of the sequence, Gilbert suggests that the Louds, unlike the Bradys who come together, will inevitably break apart. According to Eric Krueger, the split screen boxes illustrate that "the family is hopelessly atomized, its members incapable of coming together as a whole," that "everything in the film has significance only as a causal element of divorce," and that "the sequence presents the film as nothing more than a case study in nuclear-family pathology. Making division the essence of film and family, it runs contrary to the content of the film."[67] Rather than contrary, however, the sequence is better described as deconstructive, as a microcosm of the series' structure beginning with more conventional scenes of domestic life and sentimental recollection, but eventually exposing the threatening fault lines beneath the family's quotidian, at times superficial exteriors. The lines around each box represent the boundaries between each family member, whose various personalities cannot be subsumed entirely by the ideology of family togetherness advocated by the Bradys but instead look forward to their own personal future life trajectories.

This tension between individualism and familialism is indebted to the home mode, whose conventions guide the viewer to understand the problems among the Louds as a conflict of image and reality, of domestic life imagined as ideal but practiced as all too problematical. The first episode, for example, begins with a New Year's celebration, typical of the home mode's emphasis on parties, holidays, and renewal, but is overshadowed by Bill's absence, having been separated from Pat for four months. Furthermore, the narrator's exposition pointing out that this beginning is, in fact, really the ending anticipated in advance, provides the spectator a context of historical knowledge disabling any facile hope for a happy resolution.

This oscillation between the home mode's ideal system of values and the documentary's exposure of its fallacies continues in the following episodes. The second takes place during Memorial Day weekend, another holiday, and follows, in the home mode's touristic vein, Pat's visit to Lance in New York City as they frequent famous sites, eat at a restaurant, and attend an Off-Broadway play. Although Lance has already come out to his family as "homosexual," the word is never uttered and, like a patterned elimination, is replaced by euphemisms such as "spark of life," "energy," "different," "underground," and "niche." Despite the episode's focus on a reunion between mother and son, Pat's excessive drinking and smoking, as well as Lance's obvious annoyance and boredom during Pat's visit, break the home mode's illusion of good times with records of behavior usually edited out because they are inappropriately dispiriting. In the third episode, returning to Santa Barbara, extended sequences of Delilah and Michelle practicing for their dance recital, the recital itself, and the final moments of Pat and Bill's enthusiastic applause are some of the series' purest home mode moments. But these too are compromised by a scene in which Bill criticizes his children's creative aspirations, worried that they will fail to progress on a path toward financial independence.

This dialogue between the family resemblances among documentary and home modes of representation is brilliantly illustrated in the fourth episode when Pat visits her home town of Eugene, Oregon. Her present domestic situation as betrayed wife and inconsequential mother, documented by the Raymonds, is contrasted to her rich memories of her past role as the family's central figure, documented by snapshots and home movies, and reported by Pat herself in voice-over commentary. Beginning in the touristic vein, the episode opens with Merle, Pat's mother, meeting her daughter at the airport, but soon evolves into a more reminiscent mode as Pat's tour of her place of origin provides an opportunity to evaluate her current state of affairs according to past expectations that have gradually depreciated over time.

For example, during their drive from the airport, Pat and Merle stop at the site of Merle's house, in which Pat grew up and later visited regularly with her young family. Purchased by the city in 1962 to be razed for a parking lot, the house exists only as a memory, a reminder that the life Pat remembers most fondly can be returned to only through nostalgia. Lacking a referent available to the Raymonds' cameras, the house must instead be signified by home mode representations that fill in for Pat's absent past and family chronicle prior to the documentary project itself. These scenes conflate home mode, documentary, and memory as indistinguishable in their function for preserving a sense of the past from the ravages of history. Indeed, both Pat and Merle admit they are glad the house is gone so that they won't have to watch it "get decrepit," a metaphor for their feelings about the present. Like the old snapshots and Super-8 movies in which they and their families appear happy and young, their memories may preserve an idealized, pristine image of domestic bliss unsullied by material reminders suggesting otherwise. The home mode artifacts and narration by Pat only further confirm this desire: they focus on holidays, gift giving, love, cooperation, and abundance. As Pat says, it was the "first time we were all together in any settled way as a family," a golden age when there were "loads of presents and good things to eat."

To the degree that Pat's home mode representations of her past prop up the myths of *Ozzie and Harriet*—in particular a seamless image of family togetherness—without much interrogation of their patterned eliminations of contradictory content, Gilbert exploits their utopian images in a somewhat facile manner to support his thesis about familial breakdown.[68] By displaying home mode artifacts as "evidence" of the Louds' "good old days" in the 1950s in simple contrast to the evidence in his documentary of their less-than-happy times in the 1970s, he encourages his audience to respond, like Pat, with a nostalgic turn to a seemingly unproblematic past against which the degenerate present may be measured. But to compare Pat's home mode artifacts and his documentary footage as corresponding forms of evidence results in a sort of epistemological mendacity.

For example, to suggest that the intimacy the Louds had shared in Eugene would gradually dissipate in Santa Barbara, Gilbert exploits the very home mode convention that his cinema verité practice denies: direct address to the spectator, particularly when the young Loud children smile and wave at the camera in Pat's home movies. As spectators of these home movie sequences, we feel a warmer sense of intimacy with the Louds that so far we have been denied by direct cinema's impersonal gaze. When the action returns to the present day, filmed in a purely observational mode, this loss of intimacy becomes palpable, prompting us to sympathize with Pat's feelings that age and change have wedged a distance between her and the children she once used to hold on her lap, sing to, and care for in a mutual bond of love and trust. Although extremely effective as a tool to foster emotional identification, this shift speaks more to Gilbert's manipulation of technique than an actual demonstration of the family's atrophy from past to present. Indeed, when Pat's misty reminiscence subsides, so do the home mode representations, returning her and the audience almost jarringly to Gilbert's dry-eyed documentary, resuming like Merle's car, to "drive on" to the next destination, each one a new revelation both of the Louds' family history, portrayed through home mode artifacts, and of Pat's shattered goals and expectations as a wife and mother represented by the Raymonds' documentary footage in the present.

An exemplary sequence illustrating the documentary's dependency on, and exploitation of, the home mode occurs when Pat visits her old house in which she and Bill first raised their children. Because the current owners are not home, Pat cannot enter the premises, and thus the Raymonds are once again denied access to actual material signifiers of her past. Peering through the windows into a dark interior obscured from our view, Pat must rely on her memory and home mode artifacts to represent the activities she and her family once shared inside. Each window may be thought of as a page in a family album or home movie screen on which Pat projects her idealized images of the past: looking in, she wants to see not the new inhabitants but ghosts of her husband and children when they seemed to love and rely on her with greater ardor than back in Santa Barbara. A poignant moment occurs when Pat, looking lost in front of the house, is juxtaposed to an old snapshot of her standing in precisely the same spot, but smiling, one child hugging her waist, the others surrounding her as the center of their universe. When we cut back to the present, to Pat all alone, the effect is moving, indicative of her diminished sense of importance.

Again, Gilbert uses the home mode's idealized representations to suggest the series' underlying theme of unfulfilled promise. Although referring to the house itself, Pat expresses her dismay that the present generally fails to live up to her hopes for the future derived in the past: "You expect things to stay the way they are, and then they get all fouled up. . . . It hasn't been changed for the better at all." As if to confirm that genuinely good times are a thing of the past, the episode includes a birthday party for Merle that seems more alienating than communal. Although documented with conventions of the home mode, including singing, toasting, and gift giving, which should connote happiness and togetherness, Pat's unfamiliarity with Merle's guests, many of whom seem merely acquaintances to begin with, imbues the affair with a disinterested, by-the-book formality that reveals the distance, rather than the closeness, that age and geographic separation have yielded between mother and daughter. We can only wonder, doubtfully, if Pat will look back on this celebration in the seventies with the same nostalgia she has for the fifties.

After this episode's detailed contrast between the happy home mode images of the past and the deflating documentary images of the present, the series gradually moves toward a critique of social relations, within both the Loud family and their society at large, responsible for shattering Pat's illusions of domestic harmony. For example, in the sixth episode, we see for the first time real irritation and antagonism between Pat and Bill, in particular during a scene at a restaurant in which husband and wife admit that their marriage is compromised: for Bill, by his impatience at his unemployed children's lagging around the house; for Pat, by Bill's philandering. This tense confrontation is immediately followed by the paradigmatic establishing shot of their home à la domestic sitcom style, foregrounding more clearly the irony of Gilbert's thesis.

With each advancing installment, the series gradually relinquishes its earlier nods to the home mode's cultural functions of family integration in favor of the critical documentary's interrogation of family disintegration. Although the seventh episode begins in the home mode with a barbecue, Pat's slow burn at the dinner table, and her smoldering questions to Bill about his recent whereabouts, threaten to explode in a conflagration, like the brush fire that had come so close to destroying the Loud homestead the previous evening. Soon after, when Delilah telephones her boyfriend Brad to complain about her parents' constant fighting, we overhear that his parents are also undergoing the same kind of problems. Thus despite his protestations otherwise, Gilbert begins to portray the Louds as an exemplar of "the" rather than merely "an" American family: they begin to stand for a growing social trend toward familial dissatisfaction, succinctly expressed by Delilah's choice to live with Brad first before making any marital commitment.

Thus the iconography of the Nelsons as America's perfect family is replaced by the Louds as a new icon of America's dysfunctional family. For example, later in the same episode, home mode expectations are again ironically upset during Fiesta Week, an event that might inspire family fun and celebration but instead devolves into a drunken display of domestic despair. Despite the backdrop of parades, mariachi bands, and general festive behavior, Pat and Bill gradually reveal in their barroom conversations that separation is inevitable, that their marriage is a "sad comedy, that Pat thinks Bill is a "goddamn asshole." When Bill admits to a friend that "we're mad at each other all the time," his chum's reply, "You're married all the time," couldn't better express the documentary's tonal transformation from consensus to cynicism about the prospects of contemporary marriage. From this point on, episodes 8 through 12 escalate the tension and sharpen the critique with confrontational scenes rarely recorded in the home mode, if at all, particularly the notorious moment when Pat asks Bill for a divorce. For the most part, the action in the last third of the series is comprised of scenes of bickering, dissent, and condescension among the family members at each other's expense—literally in the case of Bill and Pat, who nickel-and-dime each other during legal financial settlements.

In typical cinema verité fashion, Gilbert and company end the series without providing their own explicit judgments about, or reasons for, the breakdown of the Loud family, but do allow Pat and Bill to offer explanations. Both husband and wife cite the ideology of nuclear familialism as the culprit, with Pat its unwitting victim and Bill its willful transgressor. When visiting her brother Tom, Pat realizes that her fifties image of family was a sham despite her idealized memories and home mode artifacts: "Now I understand that it hasn't been that kind of scene. . . . It was phony to begin with." Comprehending that attempts to preserve her family according to outdated and illusory conventions must fail as an anachronistic fantasy, she is empowered to file for divorce. Likewise, when Bill's inclinations toward work and extramarital affairs make the nuclear family seem oppressive, he diagnoses the ideology of familialism as a "bad bill of goods" that can only construct an unrealistic model of domestic life. In a letter to Lance explaining the cause of the divorce, Bill repudiates "a living style of all or nothing" in which "we must be together in all things," a practice to which he is unwilling to conform. Like Lance, Bill perceives the ideology of familialism as squelching his individuality, both creative and sexual, but still within a typically capitalist, patriarchal work ethic that Lance also rejects.

In summary, the success of *An American Family*'s interrogation of nuclear familialism depends to a large extent on the audience's acquaintance with the representations of family idealized in 1950s sitcoms, snapshots, and home movies as a relief or foil by which the Louds may be evaluated in the 1970s. Rather than antithetical to the home mode or to domestic television, the documentary is structured in dialogue with them to generate an idea of family as the product of intertextual signification rather than as a record of "raw reality." Yet critics such as Davis, who conceive of TV as antireality, suggest that the significance of the series is that like activist video, it countered television's "myths" in the service of "reality":

> Perhaps it was no accident that *An American Family* was revived during the very winter when George Holliday's tape in fact subverted the very networks who used it, by turning their esthetic upside down, on the lip of a decade that may yet belong to the Minicam, not the networks. Raw video is an affront to virtually all the media myths that have impacted us for more than four decades.[69]

Davis perceives the series as a precursor to amateur video because it opened up televisual content to a range of representations exceeding the typically whitewashed portrayals of network TV. Yet he ignores how *An American Family* looks not only backward at domestic television as a defining constituent but also forward to the conventions and cultural functions of domestic video, perhaps even more so than the activist video with which he aligns the series.

Perhaps the most important legacy of *An American Family* is not its politics but its revelation that ordinary people may achieve a brief moment of stardom by being broadcast nationwide. Like the Nelsons, the Louds earned both fame and criticism during the run of their series, but unlike Ozzie and Harriet, already well-known show business personalities, Bill and Pat achieved notoriety simply by airing their dirty laundry on television. Although some critics derided this pandering to audience voyeurism, as Dan Graham has pointed out, "this objection overlooked the existence of programs like *Candid Camera* or *The Newlywed Game*."[70] Here we might add that Davis's objection likewise sidesteps the existence of programs like *America's Funniest Home Videos,* a network television series dependent on reality-based video rather than its antithesis. As this popular series indicates, our curiosity about viewing ordinary families on TV may testify not just to a taste for voyeurism but to the dream of seeing our own families on television as well—a possibility no longer limited only to the

Louds or made possible only by hired documentarists, but open to any of us with a camcorder and a little luck.

An American Family points ahead to home video because it functions like a home movie broadcast on television. Today families routinely transfer their Super-8 recordings onto videotape so that rather than have to watch them on a screen, they may satisfy the desire to view themselves on a monitor, which carries connotations as a public medium transporting private moments to the world at large. According to Jay Ruby:

> Documentaries are often regarded as elaborate home movies by the people in them. Subjects become "documentary pop stars" and realize their 15 minutes of fame rather than critically examine how their images are constructed and the potential impact on audiences. The complex reaction of the Loud family to their public image in Craig Gilbert's PBS series, "An American Family," and Alan and Susan Raymond's "The Louds: An American Family Revisited," is one example of the ambivalence documentary subjects feel about their moment of ciné fame.[71]

As Ruby acknowledges, in Revisited the Louds reflect on the documentary much in the way families evaluate their representation in home mode artifacts such as wedding videos or family portraits, judging their perceived accuracy or "truth to life." They use the documentary, as if a home movie, to remember good times as well as account for the more embarrassing moments, to review the past and make sense of the present—a spectatorial process no different from Ozzie and Harriet's use of their television reruns to gauge their children's development over time, or from our perusal of home videos to laugh at ourselves, cry over lost loved ones, or simply wonder about days gone by.

►

The Wonder Years: "Dramedy" in the Home Mode

In the mid-1980s, network television producers briefly experimented with a format called "dramedy," a hybrid of the half-hour sitcom and hour-long drama that had become staples of prime-time programming. Of the several dramedies that came and went, only The Wonder Years, created by Neal Martens and Carol Black, achieved high ratings and continued popularity over the course of several seasons, while other series, such as Frank's Place (CBS) and The "Slap" Maxwell Story (ABC) barely survived a single season, perhaps because audiences seemed unsure of how to respond to their comic situations without the generic cues that they had come to expect from TV comedy, such as laugh track, one-liners, and slapstick. The Wonder Years, on the other hand, despite a similar lack of such cues and its frequently sober endings, could more readily be placed within the domestic television tradition of *Ozzie and Harriet* and *An American Family*, since it focused on the everyday activities of a white middle-class nuclear family from Southern California.[72] One might argue that its "dramedy" hybridized the comic situations of the Nelsons with the dramatic conflicts of the Louds, such that the Arnolds, the protagonists of the series, synthesized a dialectic between sitcom and documentary modes of representation, between family integration and disintegration, between the status quo of suburban sanctuary and the changing values of social revolution.

Like its predecessors, *The Wonder Years* is deeply structured by the home mode. Its serial narrative, home movie sequences, narrating voice, and plots addressing rites of passage such as attending the first day of high school, earning a driver's license, and getting married all mirror the conventions of any typical home mode collection. After the pilot's debut, reviewers immediately recognized the series' affinity to the home mode. For example, Miles Beller compared audience spectatorship of each episode to "the effect of leafing through a photo album with a good pal and seeing in those printed scenes the part of one's self now forgotten,"[73] and Tom Carson declared that "the stories don't function as stories, only as yet another snapshot to be pressed into a scrapbook, embellished by an endless, enraptured rhetoric of memory."[74] And Barry Golson concluded that "unlike the tidy moral lessons that wrapped up so many other TV family series . . . it seems to me that the wonder of *The Wonder Years* was how often life's loose ends were left dangling, as in real life."[75]

In the tradition of *Ozzie and Harriet* and *An American Family*, the credit sequence for *The Wonder Years* provides a pithy statement of its domestic themes, in this case by self-consciously inscribing home mode conventions into its structure. Simulated as home movie footage, with typical content such as sports activities, outdoor barbecues, clowning, and waving, and typical stylistic "errors," such as shaky camera movement, ellipses, and light flares, the sequence (in contrast to the slick television production methods in the 1980s) serves as a thematic metaphor for a simpler past that the series attempts to evoke. The simulated scratch lines further suggest that this home movie footage has aged: despite the present tense in which the diegesis transpires for the most part, as spectators we are to understand that all of the action has already taken place, as if the home movies of the Arnolds, restored to pristine condition, may now be watched on a TV monitor. The scratches also signify the gap between the past of the home movie scenes and the present day, between the activities of Kevin Arnold as a boy and his recollection of them as a man. Here is illustrated the fundamental drive of the series (which it shares with the home mode):

the need to narrativize one's autobiography with the help of visual aids that require, nevertheless, language to give them the proper context and deeper meaning.[76] Kevin the elder's present-day interpretation of his past, like Joe Cocker's credit sequence cover of the Beatles' "I Get By (With a Little Help from My Friends)," is a version of a previous recording reenvoiced by a contemporary through whom we screen its meanings with historical hindsight.

Although the credit sequence is typical of home mode conventions predominant in the sixties—fun and happy times—we do see momentary "lapses" during which the Arnolds appear less than perfect: Karen smacks Wayne on the patio, and Wayne punches Kevin on the lawn. When the family members become aware of the camera's presence, however, they discontinue their combative antics to smile sheepishly as if innocent of sibling rivalry. For example, Kevin and Wayne stop fighting to perform themselves as loving brothers by hugging each other (however mocking and unconvincing this show of affection may appear). This dialectic between congenial and unpleasant behavior, between the patterned eliminations of sitcoms like *Ozzie and Harriet* and their retrieval in documentaries like *An American Family*, underlines the "dramedic" structure of *The Wonder Years,* which intervenes between the poles of repression and expression to dwell precisely on the contentious family moments normally unrecorded in home mode artifacts.

Premiering in 1988 at the historical moment when home video had virtually replaced home movies as the predominant media practice of the home mode, the series was informed by the possibilities the new technology opened up to family representation. As discussed in chapter 2, videotape's lack of pressure to select and edit family activities has extended the range of recorded behaviors—not always flattering or consensual. At the same time, however, controls such as rewind and slow motion have increased the opportunity for manipulating and scrutinizing home mode recordings in more depth, in essence transforming the home mode into a mode of self-analysis.

We may observe this influence of home video technologies on the diegetic structure of *The Wonder Years* in an intriguing scene from an episode in which Kevin is anxious because he's been cheating with his brother's girlfriend, and wonders if he should break off the relationship. After Kevin's mother relates the story of how his father had jeopardized a close friendship to court her, Kevin does a double take when Mr. Arnold confirms that "it was worth it." "Did my father actually say that?" he asks. Suddenly the scene rewinds, and Mr. Arnold repeats himself in slow motion. "He did!" Kevin answers in astonishment.

In this scene, the shuttle noise appearing during rewind functions as a punctuation device dating a unique form of televisuality. As specific to videotape, its appearance within the diegesis, set during the seventies, is an anachronism. It functions, therefore, as a metadiegetic visual echo of the elder Kevin's narrating voice and concretizes his oral life review as a visual signifier of eighties nostalgia. Kevin the elder's ability to literally rewind his memories for an instant replay marks the desire of a video generation invested in capturing virtually every moment of family life on tape. The series can thus be imagined as Kevin's film-to-tape transfer of the home movies of his youth, his narration providing the audience with the exposition and dialogue normally absent from Super-8 recordings. Or more precisely, the program dramatizes the emotions typically repressed in home movies: the conflicts and anxieties lived out in between the formalized rituals of birthday parties and high school proms—the discordant moments, that is, more likely to be preserved on videotape, whose long-play duration and relative low cost have increased the size and breadth of our audiovisual memory banks over the last twenty years.

The series underscores the latent content buried beneath the idyllic portrayals of family life in classic domestic sitcoms like *Father Knows Best*, the paradigm by which Kevin the son still measures the quality of his life, but which Kevin the father can only yearn for with an ironic detachment. The program's appeal to a generation bred on both television and home movies negotiates the tensions evoked between organic memories of family life in competition with their electronic preservation in home movies and videotapes, and popular representations in domestic sitcoms circulating in syndication. In short, the specificity of the rewind effect represented in the clip points ambivalently in two directions as a node of intertextuality between home video and domestic TV. On one hand, its capacity for repetition, like the sitcom's serial nature and preservation in reruns, reflects home video's desire for continuity and the pleasure of reliving happy memories. On the other hand, its generation of audiovisual noise and distortion symbolizes the dissonance between the home movie's pursuit of ideal moments and the inevitably unpleasant actualities of everyday living, which sitcoms, video, and hindsight can deny through selective editing. This insertion of home video techniques into the domestic sitcom paradigm thus introduces cognitive dissonance into our experience of the genre.

This dissonance may also be traced to the historical transformations in domestic living arrangements brought on by social revolutions during the late sixties and early seventies (the period during which the action of *The Wonder Years* occurs) and documented in *An American Family*. As opposed to the Nelsons, who seemed immune to any society beyond their neighborhood, the safe haven of the Arnolds' suburban home becomes over the course of the series increasingly subject to infiltration by political events and moral issues threatening to impinge on and break up their nuclear unit. For example, as the Arnolds move into the seventies, the earlier credit sequence is modified to illustrate the impact of radical movements nationwide. Added to Cocker's vocals is a more discordant strain of guitar, evoking Jimi Hendrix and the counterculture zeitgeist: in effect, a harder sound for harder times. Interspersed with home mode images of the Arnolds, stock footage of deeply politicized, iconic images such as Bobby Kennedy, Martin Luther King Jr., John Lennon, hooded terrorists, black power salutes, war protests, and general civil unrest evoke the collision between residual moral values and their violent subversion.

This insertion of historical images among home mode images demonstrates the pretension of *The Wonder Years* toward dramedy, in suggesting that like the radical shift in culture itself, the domestic sitcom must follow a fundamentally new trajectory as well. On one level, the series is an apology for middle-class insulation from the most upsetting events of the last three decades and a defense of suburban life. On another level, it unmasks the central contradictions of mainstream American experience, functioning as a metaphor, personified by Kevin Arnold, for America's "'period of great self-discovery, joy and effusion, and yet at the same time a great paranoia' that match perfectly with the prevailing emotional swings of adolescence."[77] As a product of the eighties, *The Wonder Years* is thus at once nostalgic both for the naive charm and relatively unproblematic vision of domesticity in traditional sitcoms of the fifties and for the social relevance and gritty subject matter in the revisionary sitcoms of the seventies. When Kevin the elder looks back on his younger self, it's as if he were seeing a variation of Ricky Nelson through the eyes of Norman Lear.

Or possibly through the eyes of Craig Gilbert. In much the same fashion that the fourth segment of *An American Family* interrogates the ideology of familialism predominant in the fifties and early sixties by juxtaposing sentimental home mode images of the past with their less-than-ideal present-day counterparts, *The Wonder Years* appropriates the same structure in virtually every episode. For example, an episode set in 1968, a watershed date in American political consciousness, begins with a slide show presentation illustrating the friction between changing times and the postwar home mode conventions of the period documenting the ideology of American progress based on family growth and conspicuous consumption. Voiced over images of Jack and Norma Arnold's wedding, their first house, car, and baby, their second, larger house and two additional children, and finally a more recent shot of the entire family opening gifts at Christmas, Kevin the elder narrates: "When my parents started out together, everybody still listened to the Andrews Sisters, everybody was having babies, and everybody liked Ike. Everybody knew that if they just worked hard and did all the right things, a sort of paradise of family life lay ahead." Cutting to the "present" scene of the late sixties and Kevin's rebellious sister Karen, who disdains many of her parents' family values, Kevin amends his narration: "I guess paradise is a relative term." Here he plays on the word "relative" in two ways: it refers generally to different expectations held during different time periods and by different families, and it refers specifically to Karen herself, who like Lance Loud, another iconoclast, forces her family over the course of the series to alter their own expectations about domestic living, in essence changing the "picture" they have of family, in particular their own.

As a signifier that families, both real and TV, can no longer be represented as they had been in the fifties, the character of Karen serves as the primary vehicle through which social changes impinge on the Arnolds' residual family values, particularly in the ways she redefines domestic rituals. For example, at the beginning of the episode portraying Karen's unconventional wedding ceremony, Kevin narrates: "The Arnold family was undergoing a radical shift. We were in a period of readjustment, flux." The wedding marks that shift not only because Karen leaves the Arnold family to begin one of her own; more importantly, the form of the wedding itself, Karen's subversion of its status quo conventions, of how that rite of passage should look, transforms a standardized image of the home mode as well. Desiring to redefine her life trajectory beyond the traditional bride's future role as wife, mother, and homemaker, Karen resists the economic and ideological forces of social reproduction (which Kevin notes exists somewhere "between Dad's wallet and Mom's good intentions") by designing her own dress and commitment ceremony. Despite its "exotic" form, however, the wedding receives a detailed and loving home mode treatment, suggesting that Karen modifies rather than abolishes its cultural functions of consensus and continuity. A product of ambivalent times that appeared revolutionary on the surface, Karen's wedding ultimately speaks more about changing taste than radical sexual politics.

Throughout the series, matrimony often functions as a fulcrum for ideological critique. As in *An American Family, The Wonder Years* frequently interrogates the public face of marriage by exposing the private wrinkles between husband and wife. Like Bill and Pat Loud, Norma and Jack Arnold virtually perform themselves as a "happy couple" to their children, to their friends, and frequently to the camera, in an effort to sustain the image of a stable marriage despite the strain of personal conflict. One episode that foregrounds this tension begins with home movie simulations of Jack and Norma in scenes of kissing, hugging, and playing with their children. These images illustrate how the home mode may idealize memories of parents by selectively editing out the less loving moments they share with each other—an operation also typical of organic memories, in particular those of children who wish to believe in the constancy of the family unit. As the action of the episode proceeds, however, Jack and Norma argue in front of the children with a ferocity none of them has yet encountered. Soon Kevin discovers: "That was the first time I had ever seen my parents alone together." Thus once again dramatizing the dissonance between the patterned eliminations of the home mode and the unpleasant moments it generally edits from consciousness, the series also interrogates the elisions of content in fifties sitcoms like *Ozzie and Harriet,* whose portrayal of "good" parents seems almost indistinguishable from representations of many home movies during the same period.

Because *The Wonder Years* is simultaneously nostalgic for prototypical *and* revisionist domestic sitcoms, its critique of familialism straddles the Arnolds' picket fence, unsure whether to close the gate behind its visions of nuclear harmony as a phantasm of the fifties or to reopen it once more as an alternative to the familial fragmentation of the eighties. This flip-flop approach is well illustrated by an episode that explores the ideology of togetherness by dramatizing tensions between the Arnolds during their last summer vacation altogether as a family. Although none of the children express any enthusiasm for the trip—Karen deems it bourgeois, Wayne prefers to stay at home with his new girlfriend, and Kevin more or less wants to follow his siblings' suit—their parents demand that they continue an annual tradition. In their effort to rekindle family unity, Jack and Norma take the kids to Ocean City, a seaside resort where they spent their honeymoon. Unfortunately, twenty years later it has devolved into a noisy, crowded, expensive development no longer living up to their golden memories.

Against this deflated backdrop, the vacation begins badly and goes downhill from there, as the Arnolds seem unable to spend a moment together without bickering and feeling miserable. Standing alone on the beach, an exasperated Kevin says: "I couldn't figure out what had happened to the vacations we used to take, when we did things together and everybody was happy." Here the diegesis fades to a simulated home movie of a younger Arnold family frolicking on the beach, a signifier of the happier times of previous summers. At this point, the narrative seems to be critiquing home mode memories for constructing a legacy of expectations difficult to meet over time. But by the end of the episode, after the Arnolds finally negotiate their differences, however briefly, the narrative recuperates the nuclear family in a nostalgic image mimicking the previous home movie scenes. As the older Arnolds stroll down the beach, they appear to imitate, in position and demeanor, Kevin's home movie memories of the past, which he willingly joins in, as if the family were magically able to adjust themselves to the home mode's prescription for togetherness. The effect is conciliatory and redemptive, for although the vacation was for the greater part a disaster, a potentially bad memory, Kevin craves a form of remembrance that will code it as a "good time" in the future.

As this episode illustrates, *The Wonder Years* emphasizes the protagonist's retrospective relation to his past in general and to images of his younger self in particular. Like the Louds in *An American Family Revisited*, who look back on representations of themselves in Gilbert's documentary as a form of self-analysis, so too does Kevin Arnold recollect his family chronicle in an effort to understand the trajectories that have led him to his present situation. A classically split autobiographical self, Kevin as an adult attempts to reconstruct his point of view as a young boy through the wisdom of hindsight, whose ironic tone typical of the disenchanted eighties he hopes may in turn be reinfused with the hope and wonder of his childhood. This split—decidedly ambivalent—reinforces the dissonance at the heart of the series' relation to the values of its domestic sitcom predecessors. The primary tension of each episode, although dramatized externally by familial infighting, focuses more internally on conflicts within the self, on the narrator's wistful imagination that "if he knew then what he knows now," perhaps he may have been able to avoid the trials and tribulations of his youth. Such knowledge is, of course, impossible without the process of maturation, and the series illustrates that wisdom may be achieved only through painful rights of passage.

Kevin's obsession with retrospective narration belies Fredric Jameson's judgment that "memory seems to play no role in television, commercial or otherwise."[78] On the contrary, memory, both personal and popular, constitutes the very fabric of *The Wonder Years*. Indeed, Kevin's autobiography negotiates organic, home mode, and televisual memories as intertextual references to his past. A child who grew up with television, Kevin often mediates his sense of self-development through the TV programs of his youth.[79] For example, in one episode wherein he tries to make sense of his emerging attraction to girls, who seem both desirable and alien, Kevin works through his ambivalence by fantasizing himself as Captain Kirk in "Spock's Brain," a *Star Trek* episode dealing with a planet of beautiful but dangerous women. Kevin's popular memory of television thus informs his personal memories of the past with a prefab dramatic structure.

A retrospective episode that recollects previous moments from earlier seasons, a hallmark of the series, begins with the title "Looking Backward" and proceeds to mark Kevin's rites of passage with media images contemporaneous to his ongoing development, like a public sidebar of popular memories along which the audience may also locate themselves in their own personal histories. Over a montage of black-and-white images from the fifties, including JFK, Mickey Mantle, Sky King, and hula hoops, Kevin narrates: "Once upon a time, life was simple. Cars were big, gas was cheap, front yards were green. . . . The good guys were good, and the bad guys were bad. You knew where you stood." Mixed in with these historical images are both home mode images, such as family picnics, and domestic television images, in this case, appropriately, Ozzie and Harriet Nelson. Kevin's remembrance of the fifties and sixties is intermingled with idealized media images of the period, which have apparently interpenetrated his organic memories as well. "Then in 1968," he continues, "something happened, something big." Suddenly the montage cuts to an image of an exploding atom bomb, a symbol of the revolutionary changes in store for the seventies, but personalized by Kevin's own particular rite of passage: "I turned twelve years old and entered junior high school." Here the narration appropriates popular memory of macrocosmic events to make sense of microcosmic autobiographical identity, providing the voice and structure not only for the episode but for the series as well. As if in reaction to the elitist condescension of critics like Jameson, Kevin concludes his recollection at the end of the program in an uncharacteristically acerbic tone: "I think about those days again and again, whenever some blowhard starts talking about the anonymity of the suburbs or the mindlessness of the TV generation, because I know I'll never forget those times, those years of wonder."

By embedding scenes from previous episodes into later ones, *The Wonder Years* suggests that television reruns function as popular memories with personal valence. In another retrospective episode, Kevin, caught flirting in a burger joint by his girlfriend Winnie, attempts to win her back by trying to recall memories of their past arguments from earlier seasons to see if he has learned any lessons to draw upon. Although the episode's diegesis transpires temporally in the "present," each recollected scene, in effect a rerun, is presented as a flashback. Thus the series not only displays its own internal memory of itself but requires audience recollection of each episode as well to make sense of Kevin's epiphanies. At the same time, individual viewers may be inspired to reflect on past scenes from their own autobiographies, stimulated by the images of the late sixties and early seventies that *The Wonder Years* stockpiles as an archive of American popular consciousness:

In our media-dominated society reading stimulates mental images that derive as much from our photographic as our eyewitness experience. Almost every word we read refers to a concept, and for every concept there is a particular that falls under it or instantiates it, and interestingly the particulars that instantiate many concepts derive from photographic or electronic imagery rather than direct experience.[80]

Our sense of the real world, that is, may really be its mediation.

▶ ────────────────────────────────

The Real World: Reality TV in the Home Mode

By the late eighties, "reality-based" media entertainment, from *Real People* to *Cops,* had ensconced itself as a staple of television programming. In part, this new genre arose because of the increasing ubiquity of video in everyday life, including surveillance, amateur, and home video practices that recorded events and behaviors previously immune from mediation. Mainstream television soon enough appropriated these video texts for broadcast, coding their aesthetic qualities as "real" in opposition to fictional genres such as the domestic sitcom. But as these new series exhausted material and ideas, network executives concocted a variety of hybrids that exploited both video's imagined realism and the sitcom's typical domestic themes, and that in many cases were structured by the conventions and functions of the home mode. In one programming strategy, the primary trope of *The Wonder Years*—recollection of past images of everyday life for the purposes of humor and self-analysis—was grafted onto either nonfictional texts, as in *America's Funniest Home Videos,* or fictional texts interpreted via hindsight as home mode artifacts, as in *Before They Were Stars.* In both types of program, oppositions between TV/video and fiction/ nonfiction collapse, for the reality they evoke has more to do with the memory of events than the events themselves, thus inscribing "real life" into the realm of desire, reminiscence, and nostalgia. The authenticity of the reality that these programs is based on can therefore be measured only by subjective interpretation rather than objective proof. In short, documentary, fiction, and the home mode merge into a vision of domesticity that only memory and emotion can activate and validate as "true to life."

The Real World represents a landmark in hybrid programming of this nature.[81] When Lauren Corrao, vice president of development at MTV, contacted series producers Mary-Ellis Bunim and Jonathan Murray, they were working on *American Families,* a short-lived Fox series (influenced by *An American Family*) that profiled families in crisis.[82] In a sense, *The Real World* may be thought of as a synthesis of Gilbert's documentary and *The*

Wonder Years: on one hand, it documents the nonfictional everyday activities and internal dynamics of a group of people living under one roof; on the other, this "family" is played out by hired "actors" performing their roles for the camera; and overall, its primary action focuses on recollecting and analyzing domestic relationships to learn life lessons.

A primary difference between *The Real World* and its predecessors is that the recollection of its subjects occurs closer in time to the recorded events themselves. In both *An American Family* and *The Wonder Years,* a great temporal distance transpires between action and reaction: the Louds look back at themselves only in *Revisited* a decade later, and Kevin Arnold remembers his childhood only after becoming an adult and father himself. This difference of nearly synchronous from asynchronous analysis of past events parallels the difference of film's delay between production and reception from video's closure of this gap. *The Real World* thus demonstrates video's ability to telescope cinema verité's distance between producer and subject:

> The difference between this show and a documentary is that we're not going at it with a certain point of view. . . . We're letting them tell the story and the focus is on their relationships. In that way it's like a soap opera because it's about interpersonal relationships between men, women, and friends. With a documentary you try to tell a specific story.[83]

Because the series is concerned more with the aftermath of events than with the events themselves, more with narration than with narrative, a video aesthetic lends itself more readily than film to rapid self-analysis. A good illustration occurs in the second season, after David is accused by Tami of attempted rape. As David wishes he could review the incident to prove his innocence of malicious intent, the action cuts to the video footage of the event in question. Played back in black-and-white to evoke the flashback conventions of cinema, the video also functions as visible "evidence" accessible to all of the housemates for their perusal and judgment. Its recorded behaviors, of course, are subject to a range of interpretations. For instance, one might read the extended footage as proof of David's persistent aggression despite Tami's protestations to stop pulling off her bedcovers. Or from another perspective, Tami's laughter might signify that she understood David was only kidding around, thus compromising the solemnity of her accusation. In either case, the "objective" image tells only part of the story, which must be supplemented by the subtext of the participants' subjective memories of their emotional states at the time.

If *The Real World* illustrates a move from the home movie aesthetic of *Ozzie and Harriet, An American Family,* and *The Wonder Years* to the aesthetic of home video, it also marks the transition from its predecessors' nuclear families to families we choose as this decade's predominant domestic arrangement.

> A generation ago Ozzie, Harriet, David, and Ricky Nelson epitomized the American family. Over 70 percent of all American households in 1960 were like the Nelsons: made up of dad the breadwinner, mom the homemaker, and their children. Today, less than three decades later, "traditional" families . . . account for less than 15 percent of the nation's households. As American families have changed, the image of the family portrayed on television has changed accordingly.[84]

The first episode of the third season illustrates the tensions between the residual nuclear family and emerging families we choose. The introductions of cast members Cory and Pedro dramatize their transitions from their biological families to a socially constructed home. In typical home mode fashion, Cory is recorded saying grace over a final family meal and hugging her father good-bye at a train station in Orange County; Pedro's family in Miami give him a going-away party, including a cake and a chorus of "For He's a Jolly Good Fellow," all documented by a relative on home video. When they arrive at the house in San Francisco where they meet the rest of their mates, all must lay out a new set of discursive rules by which to live based on social efficacy rather than naturalized kinship hierarchies. Articulating the self-conscious irony of forming this ad hoc family in the absence of parents, Judd's first words upon entering the front door are "Mom, we're home." And to break the ice and effect their rite of passage from strangers to family members, the housemates immediately proceed to take Polaroids of each other: a mode of media practice that turns a mere house into a home.

This "acting out" as a family carries the theatricality of domestic life in series such as *Ozzie and Harriet* into a new dimension. Rather than strictly a "family we choose," the housemates in *The Real World* are more accurately its commercial simulation, a "family we audition for." Selected for their idiosyncratic personality traits and sociocultural backgrounds, the "cast members" are encouraged to play out the problematic consequences that their various ethnicities, religions, politics, and sexual orientations might precipitate when catalyzed in the close proximity of a shared domestic space. Admits producer Bunim, "We designed this family. And then we put them together to see what might happen, hoping for a dramatic effect."[85] According to these intentions, we might see *The Real World* as an inversion of the home mode's patterned eliminations, or as the logic of *An American Family* taken to an extreme: although everyday family living is

documented, it emphasizes friction rather than togetherness. As the introductory refrain forewarns: "This is the story of seven strangers picked to
live in a house and have their lives taped to see what happens when people
stop being polite and start getting real." The incidental tensions of Gilbert's
documentary are programmed into the fundamental structure of Bunim's
series: whereas the Louds had to negotiate the tension between an ideal
image of family and the oppressive roles it demanded, hoping to find a
resolution, *The Real World* begins with the expectation that modern domestic living is by nature volatile and subject to discord, such that audience
interest lies not so much in the solution to conflicts but in the behaviors
and attitudes that cause them to disrupt the household.

For all its subversive pretensions, however, the series does retain patterned eliminations reminiscent of the home mode and conventional domestic television. Despite the focus on contention, actual scenes of hostility
and sadness are broadcast to a minimum, replaced instead by rap sessions
occurring after the fact, during which housemates discuss the conflicts more
rationally in gestures of cooperation. And during these sessions, obscene
language is censored by bleeps according to broadcast standards more in
line with family sitcoms than critical documentary. Furthermore, as if a
throwback to *Ozzie and Harriet,* financial problems rarely impinge on the
social worlds of the housemates, as MTV executives provide them rent,
food, and spending money.

More significantly than its patterned eliminations, however, *The Real
World* appeals to the home mode's function of consensus to bridge difference and reconcile acts of estrangement. In the third season, when Cory
meets with Puck for the first time after he's been voted out of the house,
her excuse that reunites them is to bring him snapshots of their memorable
travels together, attempting to heal the rupture in their friendship by exploiting the home mode's ideology of togetherness. And when Pedro feels
caught between his blood ties to his biological family in Miami and his
common interests with his discursive family in San Francisco, unsure with
which to spend the rest of his life clipped short by AIDS, this tension is illustrated by the album of photos he has collected of his newfound com-
rades, formal evidence that his familial affections have been split on each
coast. By the end of the season, as the cast become increasingly nostalgic
about their seven months spent together, Judd stages a variation of *This Is
Your Life* for Pam on her birthday, encouraging each housemate to perform
for her a funny memory of good times. Adopting the recollective form of
both the home mode and a television series from the past about the past,
the sequence illustrates how theatricality, intertextuality, and popular
memory construct contemporary families through a process of mediation.

At times this process is worked out through the video documentary itself, a mode of observation that transforms into the home mode whenever the housemates invest the footage with domestic functions. For example, in the second season, during an Outward Bound excursion, the group members discover a mutual dissatisfaction with their guides and bond together not merely as friends but as family. Jon speaks for the rest when he says: "This is the first time that I can actually recall that we all really stuck together. You know it's kind of like with brothers and sisters—you don't get along with them but you know if someone messes with them, you'll get together." Here the video documentary about seven strangers functions like a home video about seven siblings. During the excursion, Irene begins to feel nostalgic in advance: "There was a time when I was walking down the trail by myself where I wanted to cry because I was going to miss everybody." As if to punctuate her statement and visualize her sentiments, the scene cuts to a typical group configuration as the video freeze-frames in simulation of a snapshot accompanied by a shutter release sound effect. Rather than only a stimulus of conflict, the Outward Bound experience draws the housemates together, its final hours videotaped like a family vacation to provide an opportunity to relive it as a fond memory. By the next episode, in fact, Tami recollects the excursion back at the house, her narration punctuated by clips from the previous episode cast in sepia tones to evoke the conventions of old family albums. As in *The Wonder Years*, events documented from earlier broadcasts function like internal memories; by the time of *The Real World Reunion*, a special inviting cast members from several seasons to reminisce about their experiences, these clips function as full-fledged home videos of the past.

The Real World suggests that at least for those generations bred on television and video cameras, reality is difficult to extricate from its mediation. The ongoing success of the series has spawned a host of specials taking off from this premise. One particularly intriguing example is "Off-Camera with Dean Cain,"[86] in which the eponymous host profiles a variety of young celebrities, such as Dan Cortese, Andrew Shue, and Matthew Fox, "off-camera"—that is, how they spend their leisure time off the set of the professional television productions on which they're currently working. The program's theme song introduces the premise:

> Lights! Camera! Wait, shut it off! The run-of-the-mill stuff is just too soft. . . .
> When the director says cut that's where we start. . . . From the roving eye that
> you can't escape, the only limitation's what we can get on tape. . . . What do
> you do when you're living off camera? What do you do when the spotlight's
> off you? What do you do with no one else around? What do you do when nobody's watching?

Of course, to reveal to the television audience what these performers actually do "off" camera, they must be recorded "on" camera so that their nonprofessional activities may be broadcast. Furthermore, aware of being taped, the celebrities self-consciously perform their nonprofessional identities, thereby recuperating their leisure time as yet another aspect of their professionalism, and confirming what the theme song implies: that in our media-saturated culture, we all live as if we may be recorded at any time.

To negotiate this contradiction, or at least to mark a distinction between professional and leisure performances, Cain's strategy is to adopt (or, more accurately, simulate) home mode conventions to connote the "absence" of professional television, including frequent direct address, handheld camera, domestic settings, and audiovisual references to the past. As discussed in chapter 3, this distinction exploits residual binaries between the imagined ontologies of TV and video: the former as professional, posed, artificial, and manipulated; the latter as amateur, candid, real, and unexpected. Like *The Real World*, however, whose style the program imitates, Cain's project blurs these oppositions in the process of reiterating them. For instance, at the end of the program, Cain edits back in what can only be called a series of "outtakes," or moments caught on camera that would appear as "mistakes" marring the fluidity of his documentary. For a project committed to documenting the "real" or "candid" lives of professional celebrities, these editorial decisions suggest that even their leisure activities have been planned out in advance, scripted, and performed. Ultimately the program conveys that "off" camera behavior is unavailable to scrutiny outside of media signification.

This link between everyday behavior off and on camera, between a sense of reality and its mediation, between our lives and the lives of celebrities, has been bridged by the popular memory of domestic television programs in another subgenre of reality-based programming, exemplified by specials such as "TV's All-Time Funniest Sitcom Weddings," "TV's All-Time Funniest Christmas Moments," and "America's Funniest Friends and Neighbors." In these programs, audiences are encouraged to identify the quotidian behavior and human relations in their lives with their humorous, exaggerated counterparts in famous highlights from domestic sitcoms. The reality on which these specials has been based, therefore, is derived from the audience's direct experiences, which they compare with mediated depictions, in the process transforming fictional texts into simulated, hyperbolic home mode artifacts. For example, in "TV Laughs at Life," host Brett Butler observes that "TV has managed to deal with every major activity a person can have in his or her life: birth, death, and all that stuff in between."[87] Structured according to major rites of passage, the program ac-

cumulates classic TV moments in the order of the human life cycle, introduced by title cards such as "Having Babies," "Love and Marriage," "The
Golden Years," and "A Funny Thing Happened on the Way to the Funeral."
Butler frequently intervenes to demonstrate how much television appears
to reflect our everyday lives, and she even jests about how our own lives
seem confusingly intertwined with TV: "Like all of America, I lived my life
tuned into television. I got pregnant with Lucy, I went through menopause
with Maude, which is unusual because I don't have children and I'm not
forty yet." Joking aside, Butler acknowledges domestic television's powerful
source of identification, recognizing that personal lives and public media
texts have become almost indistinguishable for a TV generation accustomed
to being in front of both a video screen and a lens. Later she quips: "More
and more, TV is present at the first moment of our lives. Like, you can do
basic Lamaze and basic camcorder at the Learning Annex."

As Butler suggests, real life, domestic TV, and home video have constructed an a intertext mediating private moments and their public display.
Yet another reality-based subgenre exploits these interconnections by
broadcasting nationwide rites of passage generally reserved for a small,
private circle of participants. For example, "The World's Funniest Wedding
Disasters" celebrates "tapes of the worst moments of the happiest days of
your life" by exposing the humorous ironies resulting when typically utopian domestic rituals, in particular weddings, fail to proceed as planned.[88]
Voting for the couple whose tape demonstrates the most disastrous marriage ceremony, the audience awards them a second honeymoon and a
camcorder to tape it; as if offering the prizewinners freelance work, the
host even suggests that with their bad luck, perhaps another TV special
could be made from the footage.

Another special, "Will You Marry Me?" hybridizes reality-based series such as *Candid Camera* and *This Is Your Life* by setting up unsuspect-
ing women for odd marriage proposals recorded on video for TV broadcast.[89] Its texts function as both voyeuristic entertainment for the general
home audience and home mode memories for the particular couple and
their families. An odd coproduction between amateur participants and professional crews, the program blurs residual oppositions between home
video and commercial TV. Even the televisual mise-en-scène combines the
typical diegeses of wedding videos and live TV to merge the private and
public spaces of home and studio audiences. Sitting at round tables deco-
rated to evoke a wedding reception, the spectators watch the man's proposal on a large screen, on whose inset the woman proposed to is broadcast,
usually crying. After each video proposal, the host interviews the couple
about how they met, providing subtext similar to home mode narration.

Intermixed with these studio presentations are videos of unusual wedding proposals, some produced in the home mode, and others, even more intriguing, produced initially for more instrumental purposes, but ultimately transformed into home mode artifacts by the nature of their personal function for the engaged couple. For example, during a live newscast, an anchor surprised on air by her boyfriend with a ring breaks into tears, blurring her public and private identities, becoming a breaking human-interest story herself. In another example, a young woman who believes she is performing in the production of a local commercial for her family's restaurant is surprised to find it is actually a document of her boyfriend's proposal of marriage: the video thus at once functions as advertisement, TV entertainment, and home mode memory. In both examples, the private meanings of home video inflect the public forms of television.

This process of transformation by which home mode functions privatize public texts as personal memories for their invested participants motivates the impulse for reunion specials, in which cast members of defunct domestic television series gather to reminisce about their lives together on and off the set during the tenure of the series' production. Reflecting on selected clips from various episodes, the cast evaluates the diegesis less as fictional narrative and more as a documentary of their lives together as a "family," as a record of their personal and physical development over the course of each season, and as a historical marker of their identities at the time. A fine example is "Brady Bunch Home Movies," in which the cast of the highly popular idealized-family sitcom narrate the subtext behind the scenes to portray themselves as real people rather than characters.[90] At the beginning of the program, after the Bradys are described as a "perfect family," Florence Henderson, who played Mrs. Brady, addresses the camera: "Sure, that's what everybody saw. But there's a whole lotta stuff you didn't see. And this is the real story by the people who played the Bradys told by the Brady cast themselves."

The unique twist that distinguishes this reunion special from most others is that in addition to clips from the series, Super-8 film footage produced by the cast members with cameras given to them as a gift by Robert Reed, who played Mr. Brady, provides amateur images of the performers off the set. Although these home movies, like the individual interviews of each cast member narrating the back story of their professional experiences, are supposed to offer "real" exposition in opposition to the "manufactured" image of the Bradys, they actually tell us little more than the series itself. In fact, so similar to the series, they are intercut randomly with the episode's clips, illustrating the intertextuality of both domestic TV and the home mode to construct memories of the past. This intertextuality, while attempting to extricate the real identities of the cast from their fictional characters, may actually have the opposite effect, as the audience's popular memory of, and parasocial identification with, the Bradys may reign instead; as Susan Olson realizes, "I'll be forever Cindy." At the end of the special, the cast members sit together in a studio living room near a running film projector, suggesting that all of the texts under review have been interpreted within the home mode: TV series, home movies, and video interviews all function as recollection to reinscribe the cast's professional identities within the configuration of domestic relationships.

▶ ────────────────────────────────────

The Home Mode as Chronotope

From *The Adventures of Ozzie and Harriet* to *The Real World*, from sitcom to documentary, the home mode's cultural functions transform variegated television formats and domestic representations into a supertext whose intertextual relations transcend conventional genre categories. The family resemblances among the Nelsons, the Louds, and the Arnolds encourage us to conceive them as subsets of the same group, regardless of the value systems, diachronic social changes, and humorous or sober tones that distinguish them. Because the primary tensions linking these families have less to do with common thematic conflicts than with the dialectical push and pull between past and present, image and reality, and split autobiographical subjectivity, the time-space relations of chronotope rather than the iconographic inventories and spectatorial cues of genre provide a more useful analytical paradigm by which to compare individual home video and domestic television texts crossing historical periods, media apparatuses, and genres.

In *The Dialogic Imagination*, Bakhtin coined the term "chronotope," which literally means "time space" and expresses their inseparability in certain texts: time thickens, takes on flesh, and becomes artistically visible; space becomes charged and responsive to the movements of time, plot, and history.[91] Mediating between the historical and the artistic, the chronotope provides environments that limit narrative possibility, shape characterization, and mold a discursive simulacrum of life and the world.[92] In particular, Bakhtin cites two types of chronotope pertinent to a discussion of the home mode: (1) the "idyllic" chronotope, which grafts onto a specific home location the basic stages of the human life cycle, such as birth, marriage, and death, ordered in generational sequences uniting individual lives with the cyclical rhythm of time; and (2) the "family novel" (or "novel of generations") in which the idyllic chronotope's unity of place is limited to the family town house, and the primary drama concerns the hero's definition of the family circle.[93]

Like the family trees, crests, and portraits typical of traditional family folklore, these idyllic and familial literary chronotopes may be conceived as predecessors of the twentieth century's home mode media practices. If thought of as a chronotope itself characterized by its own unique time-space relations, the home mode may broadly serve as an umbrella term categorizing heterogeneous audiovisual texts that nevertheless share its cultural functions, exploit its conventions, and resemble its domestic representations. Unlike a genre, whose texts each reiterate the same set of characteristics that nominate it for inclusion in the genre, chronotopic texts refer to each other fundamentally at the level of modality but incompletely at the level of semantics. If all generic texts are brothers, chronotopic texts are sisters, fathers, and mothers as well: their family resemblances are neither identical nor exclusive to one set of features.

Victor Turner evokes this asymmetrical, intertextual process of making unified meaning among various genres with a hall-of-mirrors metaphor:

> In a complex culture it might be possible to regard the ensemble of performative and narrative genres, active and acting modalities of expressive culture as a hall of mirrors, or better magic mirrors . . . in which social problems, issues, and crises . . . are reflected as diverse images, transformed, evaluated, or diagnosed in works typical of each genre, then shifted to another genre better able to scrutinize certain of their aspects, until many facets of the problem have been illuminated and made accessible to conscious remedial action. In this hall of mirrors the reflections are multiple, some magnifying, some diminishing, some distorting the faces peering into them, but in such a way as to provoke not merely thought, but also powerful feelings and the will to modify everyday matters in the minds of the gazers.[94]

Turner's metaphor helps us understand the relation of home video to domestic television despite its belated appearance decades after TV had become an everyday phenomenon. For example, although home video may not appear within all of the series analyzed in this chapter, its "facets" are nevertheless "illuminated" in each: as a yet unmanifested desire for a more accessible amateur medium than home movies in *Ozzie and Harriet,* as adumbrated by the diminishing patterned eliminations in *An American Family,* as informing the idea of autobiography as an audiovisual life review in *The Wonder Years,* as simulated to construct a discursive family in *The Real World,* and finally, as we will discuss shortly, epitomized in *America's Funniest Home Videos,* a hybrid synthesizing the dialectic of home video and domestic television that its predecessors have portended.

The space-time relations of the home mode chronotope shift according to the text's narrativized levels of retrospection, manifested by a focus on genealogy and generational continuity, structured by a serial aesthetic, and inflected by a nostalgia for the past. Observing and reflecting on the image of oneself and one's family, the protagonists trace their development along the private diachronic axis of the individual life course intersecting with the more universal synchronic axis of rituals and rites of passage, with which audiences may also identify by inscribing their own external autobiographies onto the time-space relations internal to the text. The chronotope's subjectivity is therefore constructed as a dialectic of private and public experience shared between actors and spectators.

Constitutive of retrospective narration, memory and projection play fundamental roles in the home mode chronotope. Negotiating among organic, recorded, and popular memories of domestic life and its mediated images, protagonists and spectators must interrogate the source of these memories: the time and place of their generation, the apparatus manufacturing and projecting them, and the intentional consciousness invested in preserving them. Because theories of genre rarely account for the protagonist's or spectator's relative subject position between an image from the past and its present source of projection, the chronotope's emphasis on time-space relations better foregrounds the constructed nature, both human and technological, of its texts.

As with genre, the chronotope evolves along a "horizon of expectation." This concept, coined by Hans-Robert Jauss, attempts to account for the dialectic of production and reception and for historical continuities and discontinuities in the reception of texts.[95] Like generic expectations, chronotopic expectations depend on memory of previous texts. The difference is that chronotopic diegeses are structured by memory, as well: their horizon advances through a dialogue between the protagonist's memories internal to an individual text and the spectator's memories, accumulating each into a serialized supertext. For example, our conception of the Arnolds in *The Wonder Years* requires personal recollections not only of their evolution over the course of several seasons but of their predecessors like the Louds and the Nelsons; of our own experiences and popular memories of the fifties, sixties, and seventies; and of the images of our own families recorded in home mode artifacts during each decade. Likewise, the temporal gap between syndicated reruns and the time of their original broadcast must be filled in by memories of the past inflected by the intervening experience of history, such that we continue to reevaluate our understanding of the texts through a screen of hindsight and, in many cases, nostalgia. Only retrospective reading activates the chronotope and renews the meanings of its supertext.

For the remainder of this chapter, I will focus on the landmark series *America's Funniest Home Videos* as a case study illustrating the televisual home mode chronotope in its most highly hybrid form. As a synthesis of reality-based programming, game show, stand-up comedy, variety show, and family entertainment, it illustrates how conventional TV genres are inadequate models by which to understand its operating logic; and as a merging of both home video and domestic television, the series demonstrates that the development of both media should be studied not only as a history of opposition, negation, appropriation, and homogenization of one by the other but also as a history of their mutually constitutive, dialogic relations, at least within the local field of domestic practice.

▶───

America's Funniest Home Videos: Commercial Television in the Home Mode

As improvements to consumer equipment have narrowed the gap in quality between "amateur" and "professional" video artifacts, institutions have reacted by adapting and adopting conventions of home video to fit those of commercial television. As John Caldwell notes: "By 1989–1990 television had a ravenous appetite for videotaped footage and seemed to welcome consumer-producers with open arms. Not only were tapes now made by nonprofessionals with little training, but stations showed off and celebrated these populist origins. The viewer was being celebrated as one of television's producers."[96] The journalistic variation of this new genre of reality-based programming, inspired by George Holliday's footage of the Rodney King beating and represented by series such as *I Witness Video,* employs home video documents of crime, accidents, and disasters as "infotainment." The domestic variation, represented most famously by *America's Funniest Home Videos,* blends the privatized experience of ritual family photography with the public, mass-marketed conventions of prime-time television.

America's Funniest Home Videos first aired on ABC as a Thanksgiving special in 1989 and debuted on 14 January 1990 as a regular series.[97] Its simple premise—to solicit and exhibit a series of humorous video clips shot by amateurs who compete for cash prizes—has had a notably enduring run, both in prime time and in syndication as reruns. Rooted generally in the genre of its comical, voyeuristic ancestors, such as *Candid Camera, TV's Bloopers and Practical Jokes,* and *Life's Most Embarrassing Moments, America's Funniest Home Videos* more particularly owes its genesis to a weekly variety show produced by the Tokyo Broadcasting Company, *Fun with Ken and Kato Chan,* that featured a segment in which viewers were invited to mail in their home video clips. Vin Di Bona, who had earlier success with other TBC properties, eventually purchased U.S. rights to the Japanese concept. As executive producer, Di Bona expanded the segment into a half-hour format to fit the ABC profile of family viewing.

Although indebted to a prevalence of reality-based programming when it debuted, such as *Unsolved Mysteries* (NBC), *Rescue 911* (CBS), and *America's Most Wanted* (Fox), *America's Funniest Home Videos* had a far greater and more immediate impact on weekly ratings than any of its predecessors or imitators. Cracking the Nielsen Top 5 after only six episodes, by March 1990 it had become the number one ranked series, temporarily unseating CBS's *60 Minutes,* a feat no other ABC program had been able to achieve in twelve years. At the series' peak of popularity, producers reported receiving close to two thousand video submissions a day.

These tapes, sorted out by screeners for broadcast approval, have to meet criteria that render them suitable for family audiences.[98] First and foremost, qualifying videos must portray funny, amazing, or unexpected events in everyday life, such as animal antics, bloopers during wedding ceremonies, and fouled plays at sporting events. Because the series emphasizes the universality and spontaneity of harmless slapstick humor, tapes that depict extreme violence, offensive conduct, and serious physical injury, or that encourage imitative behavior, are strictly forbidden. Deliberately staged videos, especially events rigged to look accidental, are also disqualified.

Once a clip is approved, its creators and performers have to sign releases for broadcast authorization. Then follows a process during which clips are adjusted for uniform quality and matched in terms of production values, embellished with sound effects and wisecracking voice-overs, organized as a montage related to a loose theme, and finally nestled into the format of the program. Each episode is first taped before a live studio audience, during which the clips are broadcast on studio monitors so that the series' producers can gauge audience reaction. After subsequent reviews of the taping, producers pass their recommendations on to the staff, who edit out the less successful moments before the program is broadcast. Although labor-intensive, this method of television production is a relative bargain, costing less per episode than the average sitcom.

Television reviewers have been somewhat puzzled by the success of *America's Funniest Home Videos,* many having panned the series as yet another illustration of the American public's increasing willingness to broadcast their most private and embarrassing moments. Several hypotheses for the series' popularity have been cited: the urge of the viewing public to get on television to secure their fifteen minutes of fame; the possibility of winning a cash prize; the all-expenses-paid weekend trip to Hollywood to attend studio tapings; the tongue-in-cheek persona of the program's original host, Bob Saget, the first performer since Arthur Godfrey to star in two concurrent, high-rated series (the other being *Full House*); the universal identification with everyday life fundamental to home movies and home video; and the sheer fun of producing television about and for oneself. The producers, however, cite the program's humor as the key to its success. Taking the "Bullwinkle approach" that provokes different kinds of laughter from both children and their parents, *America's Funniest Home Videos* not only seeks to attract a wide demographic but self-consciously mocks itself as insignificant, harmless fun.

Despite its overt lack of pretension, the series remains significant on several accounts, including its international origins and appeal. Banking on the perceived cross-cultural universality of home video productions, Di Bona had conceived of the series as international from its inception. It has been broadcast in at least seventy countries and in more than a dozen languages. Di Bona has sold the format rights to producers in other nations, at least sixteen of which have created their own versions, whereas others merely use indigenous hosts.

Most significant is the series' premise that the typical consumers of television may become its producers—that the modes of television reception and production can actually be more dialogic than unidirectional. This inversion, as well as the format's unique hybridization of genres, results in peculiar effects: the professional's commissioning of the amateur for commercial exploitation; home video's simultaneous status as folk art and mass media; the promise of reward through competition that reinflects the home mode's typically noncommercial motivations with financial incentives; the stress on comedy, which excludes the more quotidian activities often typical of home video; and finally the format's allowance for a studio audience to vote for and reward their favorite video clip, maintaining home video's folksy character, while the cash prize promotes the slapstick conventions that keep home viewers tuning in.

Comparable to *Candid Camera* as a pop phenomenon, the series has permeated the popular imaginary. Like the ubiquitous catchphrase "You're on *Candid Camera*," the jingle "America, this is you" has arguably influenced the way that the present generation thinks about home video. The difference between the two series is that the participants have moved from being the object of a media joke to its subject; that is, since *America's Funniest Home Videos*, we may all imagine being Alan Funt (the producer) rather than simply being the butt of his pranks (the consumer). *Candid Camera* demonstrated that ordinary people in everyday locations were interesting and worthy of broadcast. In this sense, it anticipated *An American*

Family by providing audiences the pleasure of watching ordinary people transgress social conventions, express anger and frustration, and expose themselves to ridicule.

Other vérité formats that influenced the makeup of *America's Funniest Home Videos* include *Entertainment Tonight* (at one time produced by Di Bona), whose demystification of stardom by exposing celebrities as ordinary human beings encouraged an inverse presentation of ordinary human beings as celebrities. The series also adopts conventions of game shows, "perennially successful in their restricted niches because they use elements of reality by employing average people as players."[99] In particular, it uses the conventions of contests such as *Let's Make a Deal* (on which staff writer Alan Thicke had worked for several years), which substitute the importance of game playing for an exploration of the conversations between the hosts and contestants, who are portrayed in a bizarre or embarrassing light.[100]

With such a "lowbrow" heritage, it is not surprising that progressive media critics have dismissed *America's Funniest Home Videos* as either inconsequential or objectionable. The most typical of their arguments condemns the series for pretending to celebrate "participatory TV," all the while in the service of maintaining the hegemony of corporate television by encouraging amateurs to shoot only in the home mode, thereby sedating the oppositional potential of a mobilized, democratic camcorder counterculture.[101] Typically, these critics reiterate binaries that monolithically oppose amateur video and commercial television, making Di Bona's series accountable to one medium or the other and deeming it "contradictory" if it shares qualities of both.

Although *America's Funniest Home Videos* in no way lives up to the utopian ideals of participatory TV, it seems premature to completely dismiss the series, which has taken one more step forward toward achieving a more democratic practice in a world of commerce as likely to squelch it altogether. It has become all too easy to exploit the stereotype of commercial television as a cannibal devouring home video, predigesting it, and coughing it back up to the audience in its own image. And by overlooking the ways that Di Bona has indeed at the most basic level required commercial TV to adapt itself to the specifications of amateur video, such critics unwittingly bolster the very power structures their discourse would seem to oppose.

Indeed, the model of commercial television as an ineluctable, all-consuming homogenizer of the audiovisual texts it broadcasts subscribes, ultimately, to the "power-of-television thesis." This proposition features three problematic components. First, it establishes an equivalence between the material effectivity of lived practices (such as the home mode) and the ideological effectivity of discursive practices (such as *America's Funniest Home Videos*). Second, it assumes that the materiality of language systems is of the same order as the materiality of everyday existence (such that what you see on the show is what you get in the home audience). And third, it argues that ordinary people model the conditions of their existence according to the internal relations of discursive representations (Di Bona's program trains people to shoot only in the home mode).[102] As Conrad Lodziak counterargues, it is misleading to attribute to TV the power of compulsion and coercion; rather, the "ideological power of television in the formation of identities is at most a power which works on ground already materially produced."[103]

It has been precisely the goal of this chapter to demonstrate the grounds for linking video and television in local fields of practice rather than opposing them always at the most global level. Thus to revise an either/or approach that deforms the hybridity of *America's Funniest Home Videos,* we may turn to Bakhtin's ideas of heteroglossia and dialogism. These concepts encourage us first to isolate the various constitutive elements of the series that maintain a relative level of autonomy instead of being wholly appropriated and transformed by one structure-in-dominance. Second, they enable us to analyze these constituents as openly heteroglossic rather than oppressively monoglossic, as precipitating a variety of effects neither necessarily predictable nor easily contained by a unitary audio-visual system of conventions and values. Third, they allow for hegemonic movement such that amateur video may have its own potential to transform commercial television: that *America's Funniest Home Videos* may have changed how audiences conceive of TV even more than how they practice home video.

For example, the program has significantly altered FCC broadcast standards. According to producer Barbara Bernstein, when the series was first proposed, executives worried that audiences would reject the substandard resolution and sound of amateur video texts, but soon discovered that these anxieties derived from their own professional expectations rather than from audiences themselves. Instead of growing impatient, spectators opened up to home video's "low" quality as a different kind of product from television's typically glossy programming.[104]

This heteroglossia of amateur and commercial elements is paralleled by an ongoing dialogue between home and television producers exploited, rather than disguised, as a source of humor and identification. Di Bona has publicly downplayed his role as executive producer: "I can't produce it. I can't make it funny. I can't make it interesting. I organize what people send in to us, and that's what my main responsibility is, just to shape the show that they create."[105] Using the submitted videotapes as "found" material, series producers apply formulas derived from a variety of conventions to provide regularity to an unpredictable range of texts culled continuously from an unknown, indefinite pool of talent. As such, when broadcast, the home video clips function as heteroglot texts: music videos, comic narratives, parodies, special effects demonstrations, promos, credit sequences, and so on.

To criticize the series for adulterating the "raw reality" of the submitted clips is thus beside the point, since from its inception, the series has never pretended to serve as documentary. Its primary intention is hardly to show "true-to-life" snapshots of the American family, but to create a playful, ironic distance between amateur video footage and the formulaic television conventions that have appropriated it. Thus while much of the audience's laughter comes from the slapstick shenanigans portrayed in the clips, just as much arises from the pleasurable friction generating their awareness of the visibility of these conventions in a parodic form. In short, the surprise of seeing amateur footage used in typically commercial formats foregrounds how television produces itself.

As the hinge figure linking both home and television producers, host Bob Saget speaks for both, most of whom remain invisible. He embodies the series as its overarching narrator, providing context, subtext, and voice-overs for each clip. Like Ozzie Nelson, Saget portrays himself as an ordinary guy despite his privilege as a professional entertainer to appear regularly on television. A stand-up comic whose material typically fails to distinguish his everyday life from his jokes, Saget identifies with his audience as a father, husband, and friend but mocks these roles in a self-reflexive fashion typical of the series' ironic attitude toward domestic life. Willing to embarrass himself on the air with "amateur" jokes, he implicitly compares himself with the amateur subjects who undergo similar humiliation in the videos.

Saget frequently calls attention to the professional and nonprofessional aspects of his persona. Professional in his work relation to the series producers and unprofessional in his identification with the home audience, Saget creates a fictional antagonism between the two groups. By constantly denigrating Di Bona and his staff on the air as incompetent, he reinforces that the amateur contributions of the audience are more important than the designs of professional executives. In one episode, a cutaway to the producers pretending to nap during a studio taping, jokes about how unnecessary they are to the show's success.

Hyperconscious of its status, function, and origin as a hybrid of home video and commercial television, *America's Funniest Home Videos* internalizes its awareness that amateurs produce its primary attractions. Countering criticism that television's perpetual flow prevents or obscures any notion on the part of audiences that its texts are produced, or have a history prior to broadcast and reception, the series emphasizes its own production processes. For example, the opening graphics are a pithy visual correlate of the series' concept. Dominating the TV screen, a camcorder lens points at the audience as its iris opens up on actual submitted video clips, which are broadcast within a graphic of a TV monitor nested within the frame. This frame-within-a-frame exhibition illustrates that although the texts have been produced as home video and later broadcast on commercial television, by sharing the same exhibition technology, they must be thought of not as one medium or the other but as both simultaneously. By reminding the audience that the camcorder apparatus is the original site of production, the logo foregrounds the source of the televised artifacts, an unusual reflex in commercial TV. At the same time, it symbolizes the desire of the home audience to turn television around on itself to see their amateur productions broadcast back to them: the home video that ordinary people generally witness on their monitors as a feedback loop from their domestic VCRs may now be viewed also as transmission from a Hollywood studio.

America's Funniest Home Videos is thus the fulfillment of the desire anticipated over the course of four decades of domestic programming: the opportunity to see images of ourselves and our families on broadcast television.

> Thanks to camcorders, we're no longer surprised to see ourselves, our families and our friends on TV. Sure, we know that what we're watching is just a home movie, not *Batman.* Yet precisely because it's just a home movie, the images of us non-heroic folks on TV help us see the tube in a new perspective. Along with VCRs, video games and the rest of the home video spectrum, camcorders demystify television—a Very Big Deal, indeed.[106]

In *Electronic Hearth,* Cecelia Tichi proposes that television has become a credential of modern life. The importance of being seen on TV, she argues, has reached so deeply into popular consciousness that our cognitive and perceptual categories have superseded the issue of what is real or unreal by what is televised or not televised: "From the 1950s onward, and well before the camcorder and VCR put large segments of the U.S. population on the television screen, various texts were revealing the ways in which being broadcast—'as seen on TV'—began to constitute a new kind of ontological state in which the self, its place, its actions are ratified and validated."[107] Following Tichi, we could write the history of television's domestic representations from the point of view of a distinctly American desire to see our families on the tube, which in the absence of a home mode technology that could make that dream real, the Nelsons, Louds, or Huxtables had to substitute for the time being. Furthermore, although domestic TV and the home mode share audiovisual conventions and cultural functions, TV's hegemony as a more "important" medium indicates that in contrast to family photographs or home movies, both of which are confined to private exhibition formats, such as albums or projection screens, families broadcast publicly on television seem somehow more important as well. Thus seeing home video artifacts on *America's Funniest Home Videos* would provide them an even greater cachet. And according to Barbara Bernstein, in fact, more than the prize money or the free trip to Hollywood, public exposure has proved to be the public's greatest motivation for submitting clips.[108]

The fundamental productive logic of *America's Funniest Home Videos* is therefore not to reify the home mode's domestic ideals, as so many critics imagine, but to produce a video interesting or funny enough to be selected for broadcast. Rather than submit clips of precious or happy moments that only celebrate family togetherness, whose dime-a-dozen ubiquity generally disqualifies them as redundant, contestants are encouraged to expand their horizons to shoot situations and events typically elided in home movies. For example, "Assignment America," a regular feature of the series, solicits tapes according to various themes, many of which have demonstrated creative impulses not strictly confined to family photography. In fact, only 1 percent of all tapes screened by the producers have been chosen for broadcast; of the remainder, their content is so diverse that the home mode is too narrow a category to contain them.

Of course, these assignments in pursuit of laughter may, as is usually the case, be criticized for lacking political content. But as Stuart Hall has cautioned, this kind of critical approach to popular culture shifts between two unacceptable poles: either total encapsulation (the culture industry) or pure autonomy (the counterculture).[109] Bernstein takes a middle position: while recognizing that the series hardly engages social issues, she suggests that viewers have other options, such as switching channels to *60 Minutes*, which airs on CBS in the same time slot. Rather than hold *America's Funniest Home Videos* accountable to newsworthy formats, she understands that like the home mode, it provides a different set of cultural functions. Most camcorder users, she acknowledges, shoot quotidian, random events without a political platform in mind, nor do they record family and friends to preserve negative memories. Even George Holliday, she points out, recorded Rodney King as a stranger: would he have been as likely to record physical abuse within his own family circle? Furthermore, she rejects arguments that ordinary people have been duped into home mode practices by false ideology, citing herself as a former broadcast journalist who has intentionally switched modes when documenting her own children.

Bernstein objects to media criticism that perceives mainstream programming as a synecdoche of television as a whole. Noting that networks have been challenged by cable, satellite transmission, VCRs, and the Internet, she suggests that media advocates should turn their focus away from criticizing commercial TV programs and toward changing federal and state regulations about public access frequencies, so that amateurs with political agendas may find greater opportunities for exhibition. She adds that because of their reliance on ratings, advertising, and widespread demographics, networks will never become a home for unstructured amateur video of all sorts, which cannot be sustained without a formula or generic expectations that broad-based audiences look forward to each week.[110]

America's Funniest Home Videos' promise of ratings delivery from this audience has been guaranteed by two regulating principles: the home mode and comedy. While by no means antithetical, their relationship in the series is, however, inflected by a self-mocking irony, whose emphasis on laughter potentially denaturalizes domestic ideologies by exposing their pretensions and contradictions. Implicit in Saget's refrain to "keep your cameras and eyes safely rolling," the tone of the series encourages responsible video practices yet wishes to deflate the sense of self-importance often attributed to domestic rituals.

Bourdieu describes this tension between humor and solemnity as a "mocking sacralization."

> The act of photographing a friend's wife in a ridiculous or even an improper posture can make people laugh all the louder because it amounts to an act of solemnization against the grain, against all the rules of good taste, an infringement of the rules of good taste which expresses and, by expressing it, reinforces, the controlled lack of control.[111]

In clips such as a bride vacuuming rice out of her bra, siblings aggressively competing for the spotlight at each other's expense, or "The Backwards Family," a reverse-motion parody of typical everyday activities, rites of passage and ritual behavior are frequently denaturalized in *America's Funniest Home Videos* by tapes foregrounding the incongruous details, family politics, and performative nature of domestic life. Indeed, a special broadcast, "*AFHV's* Guide to Parenting," was designed with the purpose to parody the idea that the series might provide a source of advice for raising children properly. Clips display accidents that might be interpreted as neglect or misbehavior, such as a mother shoveling snow on her toddler's head, a baby drinking out of a dog dish, and a boy eating gum from the bottom of his shoe. Because, however, the program had to adhere to ABC's policy of family viewing, at the end Bob Saget makes an obvious attempt to deflect accusations of irresponsibility: he invites his father onto the stage to proclaim that "the key to parenting is teaching your children love and respect." Seeking a balance between dissociation and engagement, Saget concludes that "we have absolutely no idea what we're talking about when it comes to parenting" before thanking his wife for three beautiful children.

The line that the series walks between celebrating and mocking the American family through its depictions of accidental behavior has not been without controversy. For example, *The UCLA Television Violence Monitoring Report* conducted by Jeffrey Cole at the UCLA Center for Communication Policy has cited the series for portraying scenes of decontextualized violence that may prove harmful to children. In response, Mary Conley, of the division of standards and practices at ABC, who has ensured that the violence portrayed on the series conforms to the parameters of the Children's Television Act of 1990, observes that a central tension and attraction of *America's Funniest Home Videos* is the ethical question about safe and responsible camcorder use that audiences must contemplate at the same time as they laugh.[112] She argues that unlike professional journalists who record acts of violence without attempting to prevent their potentially baleful effects in order to "get the story" in all its gory details, home videographers are discouraged, by both Bob Saget on the air and producers behind the scenes, from putting their loved ones in jeopardy. Any clips, for example, that suggest that operators should have put down the camera to intervene in situations unduly dangerous, upsetting, or humiliating have been returned to the contributor with written admonishments. Although Conley has been somewhat concerned that children may be less mature than their parents to negotiate these moral and ethical questions that the series raises, no reports of harmful copycat behavior have ever reached her desk. And at bottom line, while sympathetic to some of Cole's skepticism about television violence, she indicts his simplistic conclusions about *America's Funniest Home Videos* precisely because, as a hybrid of home video and broadcast television, the series does not position spectators only as passive receivers of representations imagined wholly outside of their own everyday domains and control over media practice.

Cole almost seems to agree: "That this program inappropriately deals with violence is the finding most likely to produce a loud 'Oh, come on!' from the readers of the report. Clearly, this is not the type of violence governmental officials and many critics are thinking of when they criticize television violence."[113] Nevertheless his study has laid the groundwork, if not the excuse, for industry intervention. Fundamental to this intervention, and a major issue of this chapter, is Cole's appeal to monolithic media theories:

> Today's video signals come not only over the air but also through cable, satellites, home video cassettes and even through video game cartridges. Even though there are a number of sources, each with different rules and obligations under the law, most people still think of anything they watch on the set as "television." . . . We approach this study aware of the fact that to most of the world it is all just "television."[114]

Without supporting this broadest of statements with any ethnographic evidence, Cole collapses all forms of media that appear on a TV monitor into a single medium, eliding their specificity so that he may issue blanket condemnations at the expense of the different cultural contexts in which the content and interpretation of these various media forms actually circulate.

Although this gross assumption may seem egregious, even media scholars who would probably distance themselves from Cole's effects studies have adopted similar rhetoric about television's monolithic specificity in their own work. For example, Richard Dienst proclaims that "in an immediate technological and spatial sense, video images never stand alone but are always strung out on lines leading back to ordering processes of commercial television."[115] While perhaps more polemical than empirical, the gist of this statement underlies much of contemporary academic discourse that situates amateur video as somehow "elsewhere" when broadcast, as if the cultural specificities of its artifacts have necessarily become either indeterminate or overdetermined technologically by television. Any sense of video's own potential for determination dissolves in the homogenizing lines of TV's inescapable resolution to consume video's distinct fields of practice into a single transmission of undifferentiated flow.

[5] *The Video-in-the-Text:*
A Phenomenology and Narratology
of Hybrid Spectatorship

Cinema and video: Cain and Abel, as Jean-Luc Godard put it. With film increasingly incorporating video, the traditional face-off that so often pits the two technologies against one another—the "direct" light of film versus the "indirect" light of video—seems finally to have dissolved. Or has it simply been replaced by a derivation of this same opposition—one that currently expresses itself in a cinematic fear and loathing of the video image, which is employed to characterise a night-marish gamut of fin-de-siecle anxieties?
:: Chris Darke, "Sibling Rivalry: Cinema and Video at War"

In 1987, at the Montreal International Festival of New Cinema and Video, Wim Wenders startled the cinema community with a self-effacing gesture: acknowledging a new generation of talented filmmakers, he turned over his prize for *Wings of Desire* (1987) to *Family Viewing* (1987), by hitherto un-known Atom Egoyan. As if by rite of passage, Wenders, a master director who had habitually interrogated the effects of manufactured images on identity through film and photography, appeared to be making way for novice Egoyan, who had just begun to explore many of the same themes, but through the emerging medium of video. Interestingly enough, in his next work, *Until the End of the World* (1992), Wenders would dramatize the Oedipal themes of *Family Viewing,* turning to video through which to mediate conflicts between father and son.

The anxieties that lurk beneath this anecdote—of the usurpation of one director by another, of father by son, of film by video—speak volumes about fears of displacement, about the status of cinema's identity itself. In an era when video technologies have infiltrated film production with a new set of practices and have increasingly been dramatized as the subject of film narratives, cinema must define and redefine its specificity both *against* video as its "other" and *through* video as its supplement. If the discourse of family resemblances often pairs video and television as sibling rivals, it just as often opposes video and film as Oedipal adversaries. Cinema's precarious claim to authority, that is, must offset an acknowledgment of the newer medium's challenge to filmic institutions with a denial of video's potency to usurp them altogether.

To understand cinema's changing relation to video, as well as its efforts to maintain hegemony as the dominant media practice at the turn of its second century, I will scrutinize discourses that have polarized these media as wholly antithetical. Arguing instead for a revised model of specificity—one that accounts for heteroglossia rather than upholding pure essence—I will then focus on home video as a textual signifier embedded within filmic narratives. This "video-in-the-text" (VIT), either represented or simulated within the frame of a cinematic diegesis, will serve throughout as a multi-faceted example of the complex effects proliferated by hybrid forms, in particular demonstrating their potential to transform the cues of classical cinema spectatorship.

After detailing the formal characteristics and general functions of this "imaginary apparatus," whose signifying codes transcend its technological base, I outline a methodology capable of explaining the spectator's perception of the VIT as an abstracted figure grounded in an existential experience of video practice. To this end, I advance a phenomenological approach that apprehends the VIT as a site negotiating this dialectical interplay of video's ideal and material aspects, which therefore revises theories of spectatorship positing transcendent or technologically determined subjectivity, and restores to audiences the degree of intentionality and consciousness of their life worlds that directs and contextualizes the interpretation of cinema texts by way of their video subtexts. Finally, I demonstrate in greater detail just how the VIT inflects interpretation, first as a narrating device transforming cinematic point of view, modes of identification, and time-space relations; second as a narrative seme organizing a set of common themes, including cognitive mapping, Oedipal conflict and resolution, the failure of interpersonal communication, the autobiographical impulse toward self-analysis, and the perils of vicarious experience.

▶

Oedipal Rivalries: Film versus Video

Although video technologies were invented fifty years after film had already prevailed as the governing media prototype, and video cameras in particular were modeled according to the ocular perspective and frame ratio of film cameras, the differences between the two media have generally been emphasized more than their similarities by both scholars and practitioners, many of whom imagine that each medium documents a divergent vision of the world. This belief may be traced in part to film's autonomous status as video's predecessor: unlike broadcast television, which videotape had been developed to complement as a tool of preservation, film's basal substrate, exhibition technologies, and aesthetic techniques had been independently conventionalized and institutionalized by the time that video offered an affordable alternative in the late 1960s. Thus taxonomies categorizing film and video generally construct their "ontologies" as polar opposites, despite their family resemblances as recording technologies.

Film's "ontology," for example, has usually been based in its chemical image process. Defenders of film have often proclaimed its superiority by citing the medium's higher contrast ratio (100:1), in-camera effects (e.g., slow motion), and simplicity of postproduction editing (e.g., splicing). Likewise, detractors of video have criticized its low contrast ratio (30:1), unintended in-camera effects (e.g., image dropout, blooming, and lag), and clumsy, time-consuming insert editing procedures (prior to the introduction of on-line digital technologies).

Video's "ontology," on the other hand, has been based in its electronic image process. Defenders of video have cited the medium's primary advantages for its near simultaneity of production and reception; its clean, flexible, and nonlinear editing technologies allowing for greater experimentation; and its potential for unlimited transmission to an indefinite number of receivers. Detractors of film deride celluloid's proclivity for scratching, burning, and fading, its horizontal and vertical picture jitter during projection, and its expensive consumption of time and finances during laboratory processing.

The important question regarding film and video ontologies remains: are they truly the effects of each technology's unique properties, or are they more likely psychological and ideological connotations associated with each medium's conventional practices? Stereotypically, film's "look" signifies mediated reality (or distance), whereas video's "look" signifies raw reality (or immediacy). These different looks mark different values, not merely neutral image-making alternatives. To prefer video or film as one's medium of choice makes a statement about one's imagined relation to the world.[1] Certainly, the causes for such intense rivalries may be traced as much to financial and institutional dictates as to aesthetic preferences. As an established industry operated by vertically integrated studios, distributors, and exhibitors, supplied by film stock, projector, and camera manufacturers, and governed by academies, unions, and guilds, the institutionalization of cinema works to ensure the medium's survival despite the advent of video.[2]

Within the discourse of family resemblances, cinema's hegemony takes on patriarchal connotations. Acting out its rival relation to video as an Oedipal conflict, cinema appeals to its "paternal" origins to establish film technique as a "unitary language," described by Bakhtin as a prior discourse with an authority already acknowledged in the past and a normalized system of signification with pretensions that set it off as a language that cannot be represented but only transmitted.[3]

> Bakhtin argued that all languages are characterized by the dialectical interplay between normativization (monoglossia) and dialectical diversification (heteroglossia). This approach provides a valuable framework for seeing the classical dominant cinema as a kind of imposed standard language backed and "underwritten" by institutional power and thus exercising hegemony over a number of divergent "dialects" such as the documentary, the militant film, and the avant-garde cinema.[4]

Video represents one more divergent dialect, one more unauthorized signified, one more subordinate set of representational codes threatening cinema's claim to authority. For example, as video's cultural diffusion has accelerated in the latter half of the twentieth century, popular magazines and trade journals have generally framed the medium in terms of its capacity to compete with film, perhaps to replace cinema altogether:

> With tape in what could be considered its developmental infancy (as opposed to the relatively advanced status of film), it is no less than inevitable that video will match, and perhaps exceed, the versatility of film very quickly.[5]

> Some exhibitors project that by the turn of the century, scrambled satellite transmission to projection video theaters will be the preferred film distribution system and that the cost considerations will drive acquisition methods toward a single medium—video.[6]

> The video image as a symbol of the end of the century? But whose century? Cinema's, of course.[7]

Video's supplemental relationship to film over the passing decades has generated hybrid forms and practices transforming modes of cinematic production and reception. For example, as a preproduction tool, video may be used to plot out storyboards, scout locations, and record rehearsals. During shooting, the video assist, by now a standard feature on the set of studio productions, allows for continuity checks, for simultaneous access to the camera image by the entire crew, and for daily reels requiring no special screening room for observation. During postproduction, film-to-tape transfers of exposed footage and on-line digital editing have been employed to increase the speed and flexibility of cutting, allow for the rapid execution of special effects, preserve and protect the original film footage, shorten the lag time between production and distribution, and reduce the costs of duplication. Finally, films on video have substantially transformed cinema reception in the form of the videocassette and DVD: not only are motion pictures more frequently viewed at home than in movie theaters, privatizing a previously public viewing environment, but to accommodate the different aspect ratios of the cinema screen and the television monitor, video often writes its own technology over the original film text, introducing effects such as panning, scanning, letterboxing, and image reduction.

Of greatest concern to this study, however, are the representations of video within filmic narratives that have proliferated over the past two decades, often serving as a key plot point, an alternative point of view, or a mode of exposition. Critical and popular films of recent years such as *American Beauty* (1999) and *The Sixth Sense* (1999) have in fact set up and delivered their narrative climaxes by exploiting a phenomenology of spectatorship unique to the video apparatus. Let us turn, then, to a detailed exploration of some of the narrative functions and themes of cinematic works in which the VIT makes an appearance.

▶

The Video-in-the-Text: An Imaginary Apparatus

The representation of video within film texts over the last twenty years has increased almost to the point of ubiquity.[8] Audiovisual displays of information presented on-screen as though generated from VCRs, projected on video monitors, or perceived through camcorder viewfinders have served as pivotal moments in works ranging from the independent art cinema of *Sex, Lies, and Videotape* (1989) to the Hollywood blockbuster *The Big Chill* (1983). In the cinema of Atom Egoyan, these displays have so consistently been used in his entire oeuvre that they have registered as the imprimatur establishing his auteur status. Therefore the iconographic value of video as a cinematic textual signifier is due for theoretical and critical analysis.

The VIT, as a synthesis of both film and video, functions as a hybrid schema, framing a portion of the cinematic diegesis from the imaginary point of view of a simulated video apparatus, but one that is simultaneously material in its evocation of the spectator's lived experience of how camcorders and VCRs actually work in everyday life. This dialectic provokes a complex phenomenology of viewing, as the audiovisual information framed for interpretation shares the specificities of video and those of cinema, its host medium. In its splintering of a unified visible field, the VIT induces perceptual dissonance, adjusting film and video to each other and changing cinema's discursive rules.

Characteristics such as direct address to the camera, unfocused, low-resolution imagery, excessive panning, unexplained ellipses, and non-narrative sequencing—taboo in Hollywood studios but unobjectionable in home video, whose associations with intimacy, ritual causality, and authenticity chafe against the spectacle, linear plot, and gloss of commercial productions—have come to form a set of codes effecting a limited transformation of classical conventions. For example, video techniques have been exploited as inventive editing devices. In *Totally F***ed Up* (1994), Gregg Araki employs the camcorder's on/off recording controls in Godardian fashion, creating electronic jump cuts to break up his narrative into random celluloid fragments. In *My Life* (1993), Bruce Joel Rubin fades out between scenes by simulating a camcorder whose power has been switched off, and Barry Levinson achieves a similar effect using video flareouts in *Jimmy Hollywood* (1994).

The formal characteristics of the VIT may easily be identified, first by its "look." Typically, the VIT's marks of difference from cinema include shaky handheld camera movement; excessive panning, zooming, and focusing; noticeable drops in resolution; video noise (flicker, blur, graininess, snow); the presence of viewfinder and monitor messages (date, LED lights, "PLAY," "RECORD") on the image track; and shots made without the benefit of lighting or microphone rigs. Second, the VIT often appears as a frame-within-a-frame, in which borders appear between the images on the depicted frame (such as a TV screen or viewfinder) and the cinema's primary mise-en-scène. Recalling Genette's "metadiegesis," or story framed within a story, these borders cue spectators to recognize that the characters in the narrative are in the presence of another, embedded media artifact, which becomes a factor in their sense-making viewing strategies.[9]

A fine example of the frame-within-a-frame configuration occurs in Atom Egoyan's *Next of Kin* (1984). In an early sequence, Peter, the protagonist, and his family attend a video therapy session to work out their domestic problems. As they sit together in a single room, under the surveillance of four video cameras, the filmic mise-en-scène fragments into four corresponding video monitors that we (as well as an attendant hidden from their view in another room) may observe singly or all at once together. At certain times during the sequence, the cinema frame merges with the borders of a single video monitor, typically narrowing our focus in Hollywood classical style to a single subject position; but at others, we must split our interest among the various monitors competing for attention. Thus our attempts to negotiate the action without necessarily finding suture with one image may either increase or decrease the spaces of our identification with Peter, his parents, and the therapist. Indeed, when Peter, looking up into one of the cameras, seems to stare out at us directly, we may be unsure if his look is at us, at the attendant he may imagine is watching him, or at his own recorded image (knowing in advance that he will view the tapes later as part of his therapy).

Egoyan's sequence illustrates the VIT's complex intertextuality. Beyond merely referring to the presence of video within the film, the VIT establishes a critical relation between the director's film camera and the video cameras he represents. In other words, the divergent dialect of video counterpoints cinema's monoglossic system of signification in a heteroglossic dialogue, thus serving as a node of metatextuality to establish a critical relation between two mediated visions of the world, often perceived to be in conflict. In that video informs and is informed by the anterior hypotext of cinema, the VIT is also hypertextual, as each medium, read through the other, gains new perspective by its seeming incongruity.

The VIT's heteroglossia therefore produces a variation of Bakhtin's "double-voiced" discourse: "an utterance that belongs, by its grammatical [syntactic] and compositional markers, to a single speaker, but actually contains mixed within it two utterances, two speech manners, two styles, two 'languages,' two semantic and axiological belief systems."[10] This dialogic relation requires that when interpreting the effects of the VIT, we treat the specificities of video that are autonomous from, yet embedded within, the specificities of cinema: "We cannot, when studying the various forms for transmitting another's speech, treat any of these forms in isolation from the means for its contextualized (dialogizing) framing—the one is indissolubly linked with the other" (340). Both polyphonic and multi-accentual, the VIT generates a dynamic plurality of polysemic, polyvalued points of view in excess of any single animating consciousness or unitary, stabilized system of sign production. As a double-voiced utterance, both video and film, the VIT betrays a relative autonomy to its technological base. That is, appropriated as a cinematic signifier, video's own system of signification may be reified, abstracted, idealized. Either a representation (the inscription of video into film's system of signification) or a simulacrum (the simulation of video by film's system of signification), the VIT often functions as a statement about cinema's own self-identity through processes of reenvoicement.

One such process is "parodic stylization," described by Bakhtin as an artistic image of one language whose intentions and ideologies collide with the language representing it (364). For example, one common theme of films featuring a VIT sequence concerns the rampant ineptitude of amateur videographers, whose emotional commitment to their dubious enterprises fails to provide them the critical distance that only the omniscient filmic narrator can put into proper perspective. In some cases, such as *To Die For* (1995), in which protagonist Suzanne Stone produces an egregious, self-serving video documentary entitled "Teens Speak Out" for a local television station in her community, the parody functions at the level of narrative. In one scene, for example, the station manager, Suzanne's boss, reviews her footage on a monitor in his office with a mocking, superior attitude with which the spectator may identify.

In other cases, such as *My Life,* the parodic stylization occurs at the level of narration. In the film's opening sequence, Bob Jones, who has recently been diagnosed with terminal cancer, attempts to videotape a legacy for his son, with whom his wife is still pregnant. As Bob sets up the apparatus, positions himself in front of the camera, and checks his monitor to ensure proper recording, his lack of familiarity with the equipment and inexperience as a videographer result in a series of audiovisual gags. For instance, he accidentally frames his feet and the leg of the tripod, miscalculates a close-up by zooming in to the side of his face, and performs awkwardly during record mode. Occurring during the narrative's introduction of the protagonist, Bob's "mistakes" foreground not only his amateurism but, more importantly, the spectator's desire for a more competent filmic narrator to widen the exposition, to "place" Bob, so to speak, within a frame that makes more conventional sense. In so doing, cinema once again implicitly puts video in its place, apprentice to film's master shot. At the same time that Bob's technical gaffes parody video practice, they inflect the narrative with a self-reflexive commentary about filmmaking practice itself. The scene's initial ambiguous eyeline matches, direct address without any explicit object or destination, on-screen screens that play with off-screen space, as well as Bob's directorial experiments and editorial choices, highlight the spectator's awareness of how the narrative's sounds and images are being produced, of the apparatuses generating them, of the difference between film and video.

The VIT, since it explicitly exposes the material aspects of both film and video media, may therefore serve as a device that defamiliarizes or denaturalizes dominant conventions.[11] Although cinematic codes may generally be taken for granted and become invisible to viewers, because the VIT literally "brackets" perception within the frame of a monitor or viewfinder, it has the capacity to bring back into the field of visibility an awareness that these codes are indeed constructs. On the other hand, if the VIT stems from character motivation, functions as a plot point, or becomes so ubiquitous as to be reconventionalized, its properties as a defamiliarizing device may in turn be neutralized and renaturalized.[12]

In a sequence from *My Life*, for example, when Bob seeks out Carol, a former childhood friend, his approach to her house, shot in first-person point of view, slightly unstable, accompanied by the sound of footsteps and camcorder controls, at first appears inexplicable because not contextualized by an establishing shot, and therefore provoking viewer curiosity about who is recording the scene. Soon enough, however, because Bob's search for advice about parenting has motivated his visit, when Carol answers the door, the film camera provides a reverse shot revealing Bob as the consciousness behind the video camera, suturing the spectator back into Bob's story and reestablishing the primacy of the cinematic diegesis in which the VIT functions more like a prop than a self-reflexive device.

Because video may be simulated entirely by cinema, the VIT functions as an imaginary apparatus: a narrative image, a semantic utterance, a dialect of expression. Having no particular "place," since its codes can be reproduced by film, the VIT as cinematic signifier may be untethered to a video apparatus altogether. Nevertheless viewers maintain the capacity to identify instances of the VIT as "video sequences" grounded by a familiarity with video technology. In this sense, the VIT operates as a dialectic of site (material apparatus) and sight (discursive representation).

Because the VIT's imaginary apparatus is rooted in conceptions of its technological base, apparatus theory tempered by concepts of soft determination may be helpful for analyzing how spectators structure their interpretation of the VIT according to their historical and practical experiences of video accumulated prior to reception.

> The advance of apparatus theory over most technological determinisms derives from its ability to raise the question of whether the technological arrangement of video discourses and institutions can be said to have some ideological or epistemological effects prior to or in addition to the historical specificity of the moments of viewing. Asked in these terms, the question retains a sense of the materiality of viewing, of the unique coordinates of each individual moment of interpretation, while still requiring an understanding of what there might be that could be common to all such acts within a given cultural formation.[13]

It is important to insist, however, that the VIT, not necessarily produced by video technology itself, is better described as an imaginary signifier of the video apparatus. Thus it complicates Baudry's application of apparatus theory that posits a technologically determined, transcendental subject position. Rather, the VIT intrudes on primary cinematic identification with the film camera, splitting the supposed transcendental subject position between two apparatuses. Shattering the illusion of the cinema screen as a reflecting mirror, the VIT reinscribes it as one of many potential frames organizing multiple sources of image projection, thus reintroducing discontinuity among images in competing celluloid and electronic registers.

Nor should the VIT be confused with Metz's psychoanalytic concept of the imaginary signifier, although his work does shed light on the VIT conceived as a set of representational codes.[14] As Stephen Heath reminds us, the shift in meaning of the term "apparatus" from a technological to a metaphysical register defines specificity not by "technico-sensoriality" but in terms of codes.[15] Thus the VIT inserts into the imaginary signifier of cinema the imaginary apparatus of video, which, grounded in its material and technological existence outside cinematic representation, is nevertheless structured by a set of psychological codes with the power to restructure the codes of cinema it shares. In short, the VIT carries with it a memory of the spectator's lived experience of video practice, a subcode that must be negotiated with the dominant codes of cinema reception.

The metaphysical or psychological nature of the VIT subcode depends in great part on the cognitive processing of its projected images. Bringing to the act of cinematic reception their basic understanding of how video technologies work as a set of reading cues or expectations, spectators perceive, distinguish, and interpret the VIT's audiovisual signals as specifically video-generated within the general context of the film text. Recalling André Bazin's analysis that photographic realism depends less on the sign's transparency than on its credibility—on the spectator's prior knowledge of how camera images are automatically produced—the VIT's plausibility as a cinematic signifier exploits a more contemporary knowledge of video's technological and formal effects.[16]

A useful text by which to demonstrate this familiarity with video's specificity in general and the imaginary of "home video" in particular is the popular television series *The Simpsons*. As a domestic sitcom, the series shares that genre's postwar American obsession with consumer products, including amateur media equipment, and the manner in which their introduction into the home soon penetrated everyday consciousness. As parody, its humorous portrayals of home video practice exploit an awareness of the incongruity between the home mode's intention toward ideal representation and the accidents of everyday life that tend to deflate those pretensions. Most significant, as a cartoon, *The Simpsons* literally illustrates that the audiovisual specificities of home video, reenvoiced entirely as animated icons, have retained an autonomous signifying power easily comprehended by audiences, regardless that the qualitative difference between the animated imagery "inside" and "outside" the simulated video frames is more imaginary than real.[17] Constituting this schema are cropping lines, record messages and lights, time registers, simulated panning, fast-forward and rewind shuttle noise, separation of audio and visual channels, image degradation, scan lines, and so on. Yet none of these formal effects have actually been generated by video technology, but instead by animation techniques. How, then, do we account for their perception and interpretation as "home video"? What theory of reception can explain the VIT's historical and imaginary existence, its material base in technology, and its imaginary representation as discourse? What methodology can negotiate the reception of both tendencies at once within a single, hybrid sign?

▶

A Phenomenology of Hybrid Spectatorship

On one hand, because the animated "home video" represented in *The Simpsons* cannot be grounded in an "objective" video apparatus, interpretation on the part of spectators must rely on subjective perception. On the other, without prior knowledge of video technology and practice, at least in a schematic form, no paradigm necessary for recognizing "home video" would exist. How, then, do we account for both the individual intentions that drive interpretation and the historical pressures and material contours already shaping cognition?

Existential phenomenology, a philosophy of consciousness that locates essence in existence and meaning in the subjective and objective experience of phenomena, provides a methodology negotiating both poles in any study of reception.[18] At least three characteristics of a phenomenological approach benefit readings of the VIT: (1) it restores active intentionality to acts of reception in reaction to the passive duplicity posited for spectators by screen theory; (2) it transcends technologically determined subject positions; and (3) most importantly, it analyzes a hybrid text first as "the thing itself" rather than after dichotomizing its elements according to polar oppositions. Beginning with a phenomenological description, the VIT should therefore be attended to as it appears immediately present to the spectator, its film and video constituents equalized rather than ordered by conventional hierarchies. Only then should follow a phenomenological interpretation that seeks out the VIT's more objective features.

A phenomenological approach circumvents the tenets of mechanical structuralism that presume a technologically determined, transcendental subject position. As Vivian Sobchack notes, "the film's material existence may be *necessarily* in its immanent celluloid, chemical emulsions, and mechanisms of cinematography and projection, but its material existence is *sufficiently* in its transcendence of its technological origins and dependencies" (171). Although her comment is directed at cinema per se, it aptly describes the VIT, whose material existence may actually be embedded in (represented or simulated by) film technology, but whose phenomenological existence projects "video." Existing "within our vision but not as our vision" (142), the VIT functions as an additive structure of interpretation that takes up relatively autonomous visions of mediated experience and negotiates their different specificities. The subject position posited by these competing visions, moreover, may not be simply reduced by theories of primary identification to a single, transcendent point of view, because the VIT's imaginary video apparatus, by displacing the unquestioned primacy of the film camera's unitary perspective on the world, dispenses with the concealing unities of suture for a movement in and out of medium awareness.

Atom Egoyan's *The Adjuster* (1992) provides an excellent example of the duality, even ambiguity, of the VIT's subject positions oscillating between media apparatuses. In one scene, Hera, a film censor, records a pornographic film with a camcorder concealed in her lap from her seat in a darkened auditorium. Because the film's visual content itself remains off-screen, bracketed from our perception, with only its sound track and light from the projection booth available to interpretation, we may imagine the action either from Hera's point of view, sutured by the primary film camera into our identification with her situation, or from the camcorder's point of view, at this point a yet unexplained accessory. Particularly because Hera's blank expression closes down opportunities for empathy with her character, our attention may soon be deflected to the camcorder in her lap, even more blank, but therefore more intriguing: what exactly is it for? The camcorder constructs an empty subject position open to all kinds of imaginative prospects for the future, including the possibility that the pornographic content, invisible to our view during the scene, might later be available to our vision in a video format. This protensive extension beyond the scene recorded by the film camera creates a temporal and spatial frisson shattering any illusion of the cinematic screen as unproblematically present to the viewer, both in Egoyan's clever elision of the secondary screen within his film's primary screen and the video camera's deferment of the secondary screen's content within the narrative. In short, the spectator's "place" within the narrative cannot be tied down to primary identification with Egoyan's camera, as the VIT constructs an alternative, if yet undefined, space beyond or outside it.

Perhaps the most important question arising from the scene is: for whom is the video recording intended? As a phenomenological mode of analysis in which spectators confront films with a purpose, intentionality mediates reception and text. The VIT functions as a figure of that mediation, as an apparatus directed by and directing a conscious perception of the world, both by characters within filmic narratives and by audiences making sense of them. For example, in the opening sequence of *I've Heard the Mermaids Singing* (1987), before we even see Polly, the narrative's protagonist, who records the recent events of her life in a video journal, her consciousness intrudes in the form of her voice, tentatively expressing her presence on the audio track: "Hello, hello . . . testing 1-2-3 . . . I guess it's on." In tandem with the credit titles announcing the crew responsible for producing the film containing her narrative, Polly's direct address foregrounds not only the constructedness of the text, both outside and inside the frame, but the conscious intentionality directing her video camera, and thus implicitly by analogy director Patricia Rozema's consciousness behind the film camera. Likewise, the spectator's consciousness directs itself to the source of Polly's disembodied voice to decipher its significance within the context of an otherwise conventional credit sequence.

Thus instead of the passive, seemingly pregiven, taken-for-granted operative vision that spectators often bring to cinema reception, the VIT requires volitional acts of perception: active, reflective, and judgmental. Like consciousness, volitional perception is never empty, waiting to be filled or imposed upon by the text, but marks a choice and sets significant boundaries for interpretation. Perception, that is, functions as a phenomenological bracketing of the contents of the text's field of experience, much like the VIT's various devices that call attention precisely to the boundaries of embedded frames on which spectators must choose to direct their hermeneutic intentions. In Atom Egoyan's *Calendar* (1993), for instance, an early scene illustrates the directionality of volitional consciousness as a drama of conflicting intentions. A commercial photographer and his wife, played by real husband and wife Egoyan and Arsinée Khanjian, travel to Armenia to record twelve churches for a calendar. In the scene, framed by the photographer's still camera, whose boundaries merge with the boundaries of the film frame, the wife shoots video of one church's surrounding environs. As she begins to converse with the guide and translator they have hired for the excursion, the photographer, an invisible presence more interested in the image of churches (and his wife) than their respective backgrounds, voices his derision at her carelessness for letting the camcorder run on in record mode, thus wasting precious batteries. His wife attempts to defend herself, insisting that she be given the chance to operate the camcorder to supplement his still camera (with her own vision), but she ultimately gives in as he demands complete control over the film and video apparatuses. The scene functions both as an analogy for the relationship between dictatorial director/objectified actress and controlling husband/passive wife and as a parable about the often competing intentionalities directing the image-making process. Negotiating this conflict, the spectator's own consciousness must mediate between husband and wife, camera and camcorder, in an act of interpretation as subjective and judgmental as the photographer himself. Thus a phenomenological approach to reception accounts for the conscious perception and directed intentionality constitutive of the spectator's volitional apprehension of the VIT's self-reflexive nature.

At the same time, a phenomenological methodology can explain how individual agency must always be informed to some degree by the "already-organized and meaningful world, a world immediately 'given' to us in specific and grounded figures that are actively 'taken up' and signified without a conscious thought" (Sobchack, 71). The life world of the spectator, that is, has been structured by a habitus that in turn structures perception and directs intentionality and interpretation in precoded (although not necessarily predetermined) ways. For example, the visual gags about home video in *The Simpsons* may be recognized *as* parodies only if spectators already have inculcated a set of norms about amateur practice whose violation may be exploited for a laugh.

Reception theory has provided two categories useful for describing and locating these modes of habituation more specifically: "interpretive communities" and "horizon of expectations." First defined by Stanley Fish, the term "interpretive communities" refers to communal strategies of interpretation that, existing prior to acts of reading, enable and delimit the operations of individual readers with a perspective that is neither neutral nor purely subjective, but interested and conventional. As members of the same community will necessarily agree because they see everything in relation to the community's presumed purposes and goals, members of different communities will therefore disagree.[19] Guiding an interpretive community's acts of reception is a "horizon of expectations,"[20] a concept advanced by Hans Robert Jauss emphasizing the basic intersubjective basis of cognition that, social in origin and existing prior to the individual, draws on cultural memories from which anticipations can be made: "The successive interpretations through which a text has been perceived becomes a 'horizon' or background that sets up assumptions about a text's meaning and thus influences its current interpretation."[21]

Depending on our attitude toward video practice incurred through membership in various interpretive communities or expectations read along various ideological horizons, our readings of the VIT will vary according to our adherence to the predisposed ideas typical of that community or horizon. For example, the use of home video in *Down and Out in Beverly Hills* (1986) has generated antithetical interpretations. In the film, Max, an aspiring adolescent filmmaker, closeted transvestite, and reluctant heir to his father's business, expresses his creativity, repressed desires, and career frustration through provocative videos portraying the oppression of his everyday life at home. Applauding the film for working through problematic issues of masculinity, journalist Chris Darke writes from the perspective of a journalist defending cinema's innovative uses of home video that open up important formal possibilities: "It is immediately reflexive and mediating, installing another image-time within the time of the film."[22] Patricia Zimmermann, however, writing from the perspective of an academic defending avant-garde media practice, derides the film's representation of home video as a signifier of the degree that ideologies of familialism have co-opted every alternative to domestic use: "If the discourse and practice of home video is not appropriated, we will all join the son in *Down and Out in Beverly Hills*, endlessly videotaping our family as a form of pseudo-participation, forever trapped within a Hollywood narrative rented for our VCR."[23]

As these disagreements indicate, the interpretation of the VIT results from an external cultural constraint rather than an internal formal technique. Systematically reorganizing this brand of reception theory as "semio-pragmatics," which emphasizes the operations of social regulation on the cognitive operations of cinema spectatorship, Roger Odin has rearticulated "horizon of expectations" as "bundle of determinations" and "interpretive communities" as "institutions." "Bundle of determinations" refers to the cultural and conventional instructions that produce meaning and affect and constitute the spectator as a social subject. "Institution" refers to structures that activate these bundles, order the text's codes into hierarchies, provide the basis for an evaluation of the text's cinematic specificity, and position the spectator's affective response.[24]

Because Odin includes the home mode among the institutions that determine how spectators interpret cinematic texts, his work is particularly relevant to this study. As a complement to Bakhtin's concepts of heteroglossia and dialogism, which enable us to conceive of the VIT as a reenvoicement encouraging formal reflexivity, Odin's concept of institution adds a functional component, such that the home mode's cultural functions (e.g., storing memories or strengthening family bonds) inflect an interpretation of the VIT (and thus of the cinematic narrative containing it) with social meanings specific to that mode: "The heterogeneity of the filmic field can be described as a structure within which each institution uses a specific filmic 'dialect' and fulfills a specific *social function.*"[25]

For example, a sequence involving home video in *Crimson Tide* (1995), although singularly brief, nevertheless provides an expository statement about the social context in which the subsequent narrative may be interpreted politically. The film begins with a mock CNN newscast relating a military crisis in Russia that will precipitate the submarine conflict comprising the greater part of the action to follow. Appearing as a live broadcast interspersed with documentary footage, this program represents the public world in which the protagonist, Hunter, plays out his professional identity as a naval officer. The newscast then cuts to a birthday party for Hunter's young daughter, representing the private world where he plays out his domestic identity as husband and father. Like shorthand, Hunter's camcorder, which he uses to record the festivities, at times from his point of view, stands out as an iconographic figure of stability, family togetherness, and good times: the social functions that the institution of home video not only connotes but tries to preserve. As Hunter happily shoots the scene, however, the LED in his viewfinder suddenly indicates low power, and soon enough the camera shuts down. Looking for a replacement battery, Hunter comes across the CNN newscast on his television, shifting his mood from domestic bliss to professional concern. Realizing that an imminent nuclear crisis will mean that he must leave his family for a secret mission, he drops the camera and prepares for departure. Thus the home video in the text catalyzes a central theme of the film: to preserve the American way of life, internal domestic relations are not powerful enough—metaphorically illustrated by the camcorder's failing batteries—but must be supplemented by external defensive operations. Because the following action predominantly dramatizes American military tactics, the cultural connotations of the home video sequence at the beginning provide a condensed "bundle of determinations," if you will, as its framing apology.

From this example, we see how Odin's concept of institution structures a phenomenological interpretation of the VIT with social functions shared by a larger community of viewers. Yet as we will discover later in our analysis of the narrative functions and themes associated with the VIT, not all of the examples cited will be, strictly, home video, but also surveillance video, industrial video, and, in some cases, television. In keeping with the unified thesis of this study, however, these nondomestic uses of video may be interpreted *as if they were* home video, which must seem like a contradiction, particularly since Odin's home mode institution, although capable of change over time, is rather invariant in relation to its competing media institutions.

Jean-Pierre Meunier, a contemporary of Odin, offers a phenomenological solution.[26] Distinguishing the fiction film from the *film-souvenir* (i.e., home movie), Meunier cites the following differences: in the former, the cinematic object is unknown in its specificity, while known in the latter; the spectator of fiction films is increasingly screen dependent for specific knowledge of what is seen, the spectator of home movies decreasingly screen dependent; in fictional consciousness, the cinematic object is perceived as "imaginary," whereas in home mode consciousness, it is perceived as "real"; the images of fiction are experienced as directly given and exist "here" in the virtual world of the diegesis, while images of home movies are experienced as "absent" or "elsewhere" in the life world of the spectator.

Where this taxonomy advances on Odin is in its flexibility. Individuals reading from within institutionalized modes may not necessarily be bound by the same bundle of determinations if their phenomenological relation to the text crosses categories. That is, our interpretation of a text is always modified by our personal and cultural knowledge of its existential position as it relates to our own. Therefore one individual's surveillance video is another's home video, one person's fiction film another's home movie, oscillating between the dependency of each on the screen for specific knowledge of the images and the fluidity between the boundaries of the screen's frame and the spectator's life world.

Two examples will illustrate Meunier's phenomenological description. In *Sliver* (1993), Zeke, the voyeuristic owner of a high-rise apartment building, installs hidden video cameras in every room of the property to survey and record every detail in the lives of the residents. Although this use of video (in particular Zeke's estrangement from his subjects) hardly conforms to the cultural functions of the home mode outlined in chapter 2, in certain instances his personal investment in, and phenomenological relation to, these surveillance tapes transforms their impersonal nature into private documents. For instance, when physically attracted to his female tenants, Zeke enters their apartments to engage in sexual intercourse under the watchful gaze of a camera, later reviewing the footage and saving his favorite moments in a special archive (akin to a family album). Therefore by crossing the threshold between his own life world and those of his tenants, in effect blurring their boundaries, he reinflects surveillance video as home video.

In *Speaking Parts* (1989), Lisa, a hotel employee obsessed by unrequited love for her coworker Lance, who moonlights as a movie extra, rents his films on videocassette to review over and again only the brief scenes in which Lance actually appears in the background but seems to take center stage in the foreground of her imagination. Because Lisa projects her private desire onto each film's fictional diegesis, and because she has external knowledge of Lance within her own life world, Lisa transforms the "imaginary" characters he plays into signifiers of Lance's existence "elsewhere," thus hoping to manifest her desires projected onto each film's narrative beyond the frames of the screen or monitor into the outside world. Thus, in this case, blurring the boundaries between fiction and autobiography, Lisa recontextualizes the home video of movie rentals within the privatized functions of the home mode itself.

Together, Odin's semio-pragmatics and Meunier's phenomenological description offer a genetic methodology that posits the subjectivity of the individual bracketed by social institutions and functional modalities. Because consciousness integrates the object of its intention within a social formation—moving between subjective and objective poles of interpretation—a reception theory based in existential phenomenology accounts for bias but remains open to change. It attends to the VIT as a thing in itself yet seeks out the structures the VIT posits about video's relation to cinema. For an application of this methodology, let us turn now to investigate some of those structures, both narratological and thematic.

▶

A Narratology of the Video-in-the-Text

In this section, we will focus on how the VIT introduces innovative narratological effects modifying classical cinematic point of view, modes of identification, and time-space relations. As a figure inserting an alternative narrative or a universe within the narrative and universe of its cinematic host, the VIT may be conceived as a "metadiegesis," whose relationship to the primary diegesis may be explanatory, thematic, or simply distracting. For example, in *Shooting Lily* (1997), a film in which the protagonist, a professional videographer, exhausts and humiliates his wife, Lily, by taping her day and night, the VIT not only causes the narrative to take place but functions as a medium to explore themes of dependency, obsession, and narcissism. During the film's delirious opening sequence, shot entirely on video through a first-person viewfinder, Lily runs from room to room during a New Year's Eve party, trying to escape from the roving eye of her husband's camcorder. Aggravated by his smothering attention and embarrassed before her guests, she erupts in anger, demands a divorce, and makes a swift exit, but not before she exclaims that if he can't figure out why she's leaving him, he should go back and look through the hours of videotape he has recorded of their relationship. And so begins the film's narrative, a retrospective retracing of a doomed marriage. The video sequences, which make up most of the film's running time, provide an implicit commentary on the couple's present situation and a metacritical space for the husband to find the necessary distance to finally understand his wife's exasperation.

The film wonderfully demonstrates the indelible connection between the points of view of character (subjective outlook) and apparatus (optical perspective). Although Edward Branigan rightly defines point of view as irreducible to the camera's position,[27] and David Bordwell protests that point of view should be discussed not as a physical object but as a personifying agent,[28] the VIT is interesting precisely because it materializes human subjectivity as an apparatus, which frequently acts as a personified character itself. Thus when it appears as a camera within the film text, the VIT is a hypothesis not merely about cinematic space but about the source of the sound and images projected and the consciousness operating the technology producing them.[29]

Because video has conventionally connoted an ideology of immediacy, and home video an ideology of domesticity, the discourse of the VIT negotiates, even blurs, the impersonal narration of cinematic fiction and the personal narration of the causal agencies operating within its diegesis as characters. In other words, in instances where cinema constructs a fictional world as if it existed autonomously, directly present to our vision, without source or origin, the VIT reports on that fictional world in the manner of a witness or participant. For example, the opening credit sequence of *True Love* (1990) shares the frame with a home video sequence portraying an engagement party for the narrative's principal protagonists. From somewhere off-screen, we hear a chorus of voices admiring and laughing at the footage, as well as signs of fast-forwarding through and pausing on the action. Soon we discern that some of these voices belong to the bride and groom to be, in-laws, and relatives or friends. Engaged ourselves by the presence of these disembodied voices commenting about their spectatorship, our own typically invisible identities as spectators become reembodied as we join this informal, private gathering of characters we will soon come to know better as their story unfolds.

If cinema functions as an omniscient frame providing narrative objectivity, the VIT functions as subjective points of view within it. At the same time, as a medium of both ocularization (what is perceived by the eye or lens) and focalization (what is known epistemologically, experienced psychologically, or believed ideologically by a character), the VIT may conflate the character's objective and subjective points of view with those of the camera and narrator. For example, *Death Watch* (1980), a film speculating about the future of broadcast television, introduces a new video technology: a miniature camera lens inserted within the human eye to record the carrier's exact vision from a first-person point of view, allowing for literal "eyewitness" coverage of events. The protagonist, who submits to the operation, is hired by a television producer to track the last days of a terminally ill writer so that audiences nationwide may experience her death vicariously through his personal vision. Beginning at the purely perceptual level, the protagonist's point of view functions at the level of ocularization, demonstrated by visual gags such as panning and black masking, which simulate his moving head and blinking eyelids, respectively. As he eventually comes to sympathize with the writer's plight after exploiting her condition without her knowledge, the protagonist's point of view shifts to focalization, both psychological as he weeps in regret of his assignment and ideological when he finally withdraws from the project altogether, blinding himself as a symbolic act of negation that reaffirms the narrator's (or filmmaker's) critical point of view deriding the inhumane consequences of surveillance and voyeurism.

Just as the protagonist, internal to the film's diegesis, moves between objective ocularization and subjective focalization, so the spectator external to the text may identify with either pole elicited by the VIT. Thus spectatorial identification oscillates between the cinematic diegesis and the videographic metadiegesis, a "double vantage" described by Rosalind Krauss as both having an experience and watching oneself have it from the outside: "The spectator will occupy two places simultaneously. One is the imaginary identification or closure with the illusion. . . . The second position is a connection to the optical machine in question, an insistent reminder of its presence, of its mechanism, of its form of constituting piecemeal the only seemingly unified spectacle."[30]

In his seminal essay on *Stagecoach* (1939), Nick Browne coined the term "spectator-in-the-text" to describe this double movement that asks viewers to be two places at once: where the camera is and with the depicted person. Browne argues that cinema's double structure of viewer/viewed throws into question any technologically determined or transcendent point of view; instead, we can identify with characters and share their points of view, regardless of optical framing.[31] Although the idea of "video-in-the-text" is indebted to Browne's work, the VIT more than doubles the structure of cinematic reception. Rather, it triangulates camera and character identification with a third look neither simply the spectator's own nor a character's: that is, the spectator may identify with the look of a metadiegetic video camera. As an apparatus mediating subjectivity, the VIT may be shared by the spectator and the subject imagined to be operating it (we identify with the consciousness behind the lens), may at other times be relatively autonomous (we do not identify with the consciousness behind the lens), and may at others be without a subject altogether (we cannot identify the source of the consciousness behind the lens, as with unmanned surveillance cameras). Metz's notion of primary cinematic identification of the camera with one's own look here becomes hybrid, as one's look is no longer imagined to be coterminous with the film camera—as one's movement between introjection (one's own consciousness) and projection (the consciousness behind the camera) is no longer merely double, but multiple.[32]

Another look at the previously discussed sequence from *My Life* will illustrate. When Bob first confronts the camera directly as he records his video journal, we are unsure whom he is addressing. Is he aware of our identities as spectators? Are we the destination of his address? If not, who is? Or is he simply talking back to himself? If so, are we supposed to identify with his position, as if we are talking back to him or to ourselves? Already the possibilities for identification are kaleidoscopic. Soon, however, some of the mystery of his address unravels: Bob is leaving a legacy to his unborn son, a subject-yet-to-be. The complexity then multiplies once more: how do we as subjects-who-are identify with an a posteriori position? Further-more, when Bob remarks to the camera, "You are about to be born," oddly, this second-person pronoun takes on third-person connotations that even Benveniste would have difficulty explaining. Our subjectivity must be split between Bob's direct address to the spectator ("you" meaning "us") and to his son ("you" meaning "him"). The oscillations between "you and I," and "you and him," shift not only in terms of person but in relations of time and space. To identify with Bob's unborn child, we must project our sub-jectivity beyond the present time and place of the scene onto an imaginary future plane where Bob will have ceased to exist and his son will have come to be. In short, to take the subject position of Bob's address literally, we must shift from the moment of Bob's act of production to that of his son's act of reception, all the while acknowledging that both moments have been subsumed within the moment of our own subject positions as receiving spectators.

A mediator of this tension, the VIT functions discursively to mark narrative "tense," defined by Gerard Genette as the inverse relation be-tween the sequence in which events occur in the story and the order in which they are recounted.[33] While at the level of story (the cinematic diege-sis) events may be conceived as occurring in strict chronological sequence, at the level of discourse (the videographic metadiegesis), events may be pre-sented in an order that deviates from this chronology. In particular, the VIT frequently acts as a technologically motivated analepsis: a representation of the past that intervenes within the present flow of the filmic narrative, often understood as a subjective memory or story-being-told.[34] Offering a new set of transitional devices, such as rewind, playback, and changes in resolution, the VIT retains yet refreshes the more conventional codes of the cinematic flashback, such as voice-over narration, dissolves, and changes in color. Not requiring these "artificial" manipulations of filmic technique to signify the transition between objective diegesis and subjective point of view, the ready-made verisimilitude of the VIT replay, grounded in video's "real-life" operations, may be easier for contemporary audiences to envision as a new set of conventions registering past events.

Because of their capacity to preserve past events, communications media have often been compared to the human brain as mnemonic systems storing information through processes of selective attention (editing) and retention (recording).[35] The important difference between organic and mechanical memories, however, distinguishes the brain and electronic media: whereas human memory has a limited capacity, constantly having to determine the relevance of incoming experience, videographic memory may theoretically increase in size indefinitely, subject only to material erosion. As we will observe, narratives representing VIT flashbacks frequently mine this distinction as a moral or ethical dilemma between "real" life and its vicarious experience.

As an audiovisual tool for recapturing the past, the VIT simulates a variety of memory functions. For example, in *My Life,* Bob Jones uses his camcorder to arouse "evocation memory," the process described by Jean Piaget for educing objects in their absence by means of an image memory.[36] Repressing memories of his childhood in Detroit as the son of working-class immigrants whose incomplete cultural assimilation embarrassed him to the degree that he escaped as a young adult to Los Angeles and legally changed his Eastern European name, Bob discovers in home video a medium through which to explore his forgotten history, fill in lost details of the past, and reconnect with important friends and family before he dies.

In other films, however, the VIT may function in reverse as a medium of repression, generating a "screen memory," defined by Freud as a substitute image that screens out a painful memory too difficult for a subject to confront.[37] In *Exotica* (1994), for example, Francis, whose wife and daughter have been brutally murdered, cannot achieve catharsis to work through his mourning as again and again, rather than confront his grief, he remembers a flash of video he had shot of them during happier times. At each moment that Francis seems on the verge of epiphany, near to understanding the tragedy, the diegesis flashes back to this videotaped scene, rewinds to the moment Francis recorded the footage (black-and-white, it simulates his vision through the viewfinder), then plays out until it pauses in freeze-frame on his wife's hand reaching out to block the lens—

an apt metaphor for Francis's psychological resistance to moving forward with his life.

Along with tense, the VIT marks narrative "duration," defined by Genette as the relationship between the time measured in the story and the time measured to tell the story.[38] Examples of duration include "slow motion," in which discourse swells the time of an event that occupies a shorter time in the story;[39] and "pause," in which time stands still in the story while description is carried on at length (see 99–106). With its handy slow-motion and pause controls, the VIT actually visualizes these narrative operations. For instance, in *Henry: Portrait of a Serial Killer* (1990), the protagonists review videotapes of their violent adventures in slow motion to savor the details; and in *Jade* (1995), homicide detectives freeze on the face of a murder suspect caught unaware by a hidden video camera in order to ascertain her guilt.

In another example, the several video montages of *My Life,* in which Bob Jones leaves advice for his son on how to play basketball, cook spaghetti, shave, and jump-start a car battery, illustrate narrative "summary," in which a long stretch of story time (the passing of Bob's final days) is condensed in a brief passage of discourse (see 94–99). Here home video's typical elisions of action between "special" or noteworthy events provide the alibi disguising the pressure toward brevity in mainstream Hollywood cinema. These elisions function as "ellipses," in which time passes in the story, while no time passes in the discourse (see 106–9). The VIT provides a number of devices to construct narrative ellipses that edit out unimportant events, telescope cause and effect, or cover vast stretches of action economically: turning the power button on and off, pausing in record mode in between takes, and fast-forwarding through footage, to name a few.

Perhaps the most common ellipsis produced by the VIT bridges action and reaction by time-shifting the act of video production to the act of reception, from one present moment to another. Again in *My Life,* Bob is frequently portrayed directly addressing his camcorder, seen through the viewfinder's perspective, thus evoking the "present" moment of production. At times the viewfinder's on-screen frame identifies this moment, but at others the frame disappears when the camera moves in for a close-up, eventually pulling back to reveal the image framed no longer by the viewfinder but by the borders of a TV monitor, a new on-screen screen identifying an act of reception, a new "present" moment recontextualizing Bob's act of production as a past event and identifying a new subject observing the footage. In one scene, this shift represents not only a temporal change but a change in Bob's relationship with his wife Gail. Having previously kept the video journal a secret, even using the camcorder as a mode of therapy to admit his fear that Gail will remarry after his death, Bob finds it easier to communicate with an apparatus than with another human being. Gail's "unauthorized" reception of his confession, discovered in Bob's absence, prompts her to confront Bob about his failure to express his feelings to her. From this point on, they share camcorder duties, participating in Bob's video project together as a sign of their new covenant of trust. Thus the time shift serves both narratological and thematic functions.

As this example illustrates, the VIT ellipsis jumps not only diachronically in time but laterally in space. Bob's confession in the Santa Monica mountains is transported to his living room in Los Angeles, bypassing the distance it has traveled between production and reception without an obvious conventional cut. Genette refers to this type of ellipsis as "paralipsis," which sidesteps a narrative element rather than a moment in time (52). Although Genette's term refers more specifically to linguistic technique, if applied to cinema, paralipsis may be appropriated to investigate the VIT's experiments with off-screen space, its on-screen framing devices sidestepping elements in the primary diegesis for dramatic effect.

For example, one elaborate sequence in *Bad Influence* (1990) brilliantly creates an original form of narrative suspense precisely by eliding the spectator's conventional sense of time and space, by exploiting the tension between ellipses of time and paralipses of space. At the film's climax, Michael, the protagonist, realizes that Alex, a drifter whose carefree amorality had initially liberated Michael from the confines of an oppressive career and arranged marriage, has ultimately been a bad influence, in fact jeopardizing Michael's future and threatening his life. Returning to his apartment to make a final break with Alex and take control of a rapidly deteriorating situation, Michael discovers his television monitor in play-ready mode, inviting him to observe a videotape clearly waiting for his arrival. Inserting the tape, Michael views the murder of his girlfriend Claire by Alex in the very same apartment.

The VIT here functions as a self-reflexive essay on the nature of cinematic suspense. Like the film's spectator, Michael is forced to watch events already recorded, but as if they are happening within the present time, and in this particular case, the present space. Thus even though the film's diegesis is posterior in time and exterior in space to the metadiegesis of the video monitor, the videotape's illusion of immediacy invites us to enter into the action as though we might intervene. Like a film director, Alex manipulates the camcorder to follow only the action he wants Michael to see; like a narrator, he constantly looks back into the camera as if checking to make sure Michael is still watching, even coaxing him to "come on down" the hall to the bedroom to witness the murder. And indeed, as Michael runs to the scene of the crime, Claire's screams emanating from the video's audio channels confuse the act of the murder and its discovery, creating a fruitless hope that Michael will be able to prevent the atrocity, a hope ironically and bitterly dashed when he encounters Claire's bloody corpse.

The suspense depends on paralipses of absence and presence, of here and now. Although the life worlds of each character and the diegeses of both film and video share the same space of Michael's apartment, that space is disjointed by temporal alterities. Thus when Michael follows Alex down the hall, they paradoxically move at the same time and in the same space, but within different mediated times and spaces. Nevertheless video's role as intermediary sidesteps the boundaries between the on-screen space of the murder and the off-screen space of its aftermath, blurring them with a clever use of sound. When Claire knocks on the door within the video's metadiegesis, Alex glances over to the door in the film's diegesis, confused about its source. Not only does this ambiguous knock add dramatic irony to Claire's question when she enters the room, "Is he here?" (referring to Michael), but it demonstrates the impossibility of locating and grounding sound effects essentially on- or off-screen.

Finally, as a complement to duration, the VIT functions equally as a discourse of narrative "frequency," defined by Genette as the relation between the number of times something occurs in the story and the number of times it is represented in the discourse (113–60). Of particular interest to the VIT is Genette's concept of "pseudo-iterative," an event or action narrated as if it happened in the story repeatedly, but whose precision of detail marks it as undeniably singular (121). Noting that mainstream narrative film utterances seemingly speak the singulative because of their rich perceptual detail and simple tense that captures and embalms the present moment, Marsha Kinder has argued that classical Hollywood cinema's expression of iterative frequency is usually explicitly or implicitly pseudo-iterative in nature.[40] Although Kinder emphasizes the subversive potential of the pseudo-iterative in Italian neorealist and feminist avant-garde films, here we will observe how the VIT appropriates the pseudo-iterative more conventionally to evoke the cyclical events of the life cycle reiterated by home video.

Although pseudo-iterative, like film utterances, because of their concrete detail, home video sequences emulate the iterative, because the rites of passage they celebrate, such as birth or marriage, while unique to each instance, universally stand for all similar celebrations. For example, in *The Family Album* (1986), a montage of silent home movies recorded from the twenties through the sixties, Allan Berliner illustrates and confirms the continuity of genealogical traditions. Dramatizing the human cycle, he structures his documentary in groups of images spanning the average lifetime.

From infants at their mothers' breasts and children on a grandma's knee, to young adults at their first proms and older couples' anniversaries, iconic moments are synchronized with independent audio tracks compiled from private tape recordings of various family gatherings such as birthday parties, holidays, and vacations. Berliner's inspired unity of autonomous visual and audio tracks, whose corresponding effects seem so natural, illustrates our culture's timeless performance of familialism across boundaries of ethnicity, geography, and technology. At the same time, Berliner's ability to recognize the universal similarities between the various sounds and images reveals a generic protocol whose shared expectations maximize domestic ideology: the family album records an individual's rites of passage in such a conventionalized way that all family albums are alike. A series of cycles whose arrangement shapes memory, secures identity, and generates tradition, these forms of family exhibition are at once historical and ubiquitous, a syntactic progression of paradigmatic moments.

Particularly in film narratives concerned with rites of passage, the VIT exploits paradigmatic moments to illustrate that a character's life is unfolding in its "proper" chronology, or stagnating in a retarded stage of development, or simply attending to everyday routines. In *My Life*, once Bob's son is born, his home video montage of the baby's activities such as eating, sleeping, smiling, and playing demonstrates both the uniqueness and typicality of babies, an implicit symbol of the life cycle's inevitable turn from birth to death. And in *Sex, Lies, and Videotape*, Graham's home video collection of tapes recording women confessing their sexual fantasies demonstrates his emotional immaturity and incapacity to nurture a real, committed relationship. Although each woman is unique, her name written on each cassette's label, the tapes themselves are exchangeable, each an iteration of the next within the solipsistic loop of Graham's narcissism.

These examples, while hardly exhaustive, should suggest the complex narratological questions that the VIT raises about cinematic point of view, identification, and time-space relations. Admittedly, this narratology has been conducted in broad strokes, where a more detailed, perhaps even book-length, study would do more justice to the issues only touched on in this section. But for the sake of economy, let us turn now from narrative functions to a survey of narrative themes.

▶

The Video-in-the-Text as Semantic Apparatus

In *S/Z*, Roland Barthes refers to "thematics" as the analysis not of words but of units of signification. Constituting these units, or "lexias," are "semic

codes" that function as indexes defining an "avenue of meaning." Each
semic code functions something like a narrative magnet, organizing a satellite of connotations into a semantic figure that can be pointed to, named,
and interpreted.[41] As a subcode of cinematic narrative, the VIT may be
conceived as one such organizing figure of phenomenological interpretation, a "semantic apparatus" that generates a common set of themes exploiting or reacting to the ideologies of video practice common to the home
mode habitus. Again, although a comprehensive compendium exceeds the
limited scope of this chapter, I will interrogate some of the more represen-
tative themes illustrating the VIT's communicative, testimonial, emotive,
and ideological functions.

Cognitive Mapping: Home and the World

The traffic jam that opens *Falling Down* (1993) could not better illustrate
the dystopian, schizophrenic vision of Los Angeles dramatized by much of
contemporary Hollywood cinema. Recalling Godard's traffic jam in *Weekend,* the images grow increasingly disjointed, meaningless, and frustrating.
But where Godard's sequence is represented in a single take external to
character focalization, director Joel Schumacher breaks up the scene in a
chaotic montage reflecting the disordered mind of a man unable to process
the overwhelming sensory input of the urban landscape: signs, noise, music,
buses, cars, trucks, children, adults, animals—all brought together inter-
minably at a standstill because of road construction whose causes cannot
be seen, let alone determined. Near the breaking point, stuck behind the
wheel, the man eventually bolts from his car, leaves it stranded on the road,
and heads for home.

Thus begins the odyssey of Bill Foster, Everyman and Nowhere Man,
who wreaks havoc in his march from downtown to his ex-wife's house in
Venice Beach, where his daughter is celebrating a birthday. Identified only
by his license plate, D-FENS, he evokes a modern-day Ulysses gone awry,
a right-wing conservative who reviles the ethnic diversity, random violence,
economic inequality, and sociopathology of L.A. Busting up a Korean gro-
cery, holding up a fast-food restaurant, terrorizing Westside golfers, and
beating up Eastside gang bangers, D-FENS proclaims to the world: "Clear
a path, I'm going home!" Like *Blade Runner* (1982), *Falling Down* exploits
the city's reputation as a decentered, mobile postindustrial sprawl without
a stable identity. Such a dislocated landscape produces schizophrenia, nostalgia, and homesickness.

In a city fragmented by territorialization, D-FENS's only defense is home itself. The nuclear family guides him as a cognitive map, providing the only structured view of the world remaining to him. Ironically, however, the home he seeks is a fiction. Finally arriving at the house, vacated by his ex-wife Beth and daughter Adele in terrified anticipation of his unlawful entry, he proceeds, in the absence of a flesh-and-blood family, to watch home videos of his daughter's previous birthdays, which, like an epiphany, reveal in a moment of intense narrative economy the resolution of the protagonist's tensions. Reliving these happy moments until their magic is broken by the tape's record of his flaring temper when wife and daughter fail to perform as scripted per home mode conventions, the protagonist suddenly recognizes his own internal contradictions and understands why his wife divorced him.

Here home video documents in seconds the contradictions the commercial film has belabored for two hours: the friction between ideal domestic representation and the dysfunctional realities of modern family life. Rather than *home,* we might say that *home video* is really what he's been searching for: it functions as the narrative's primary hermeneutic code, the final answer to his enigmatic behavior. Here the specificities of the home video apparatus reveal the cognitive dissonance of patriarchy in crisis. Reviewing himself in the mirror of the television monitor, as the internal focalization of memory becomes the external ocularization of self-analysis, the protagonist must admit that father hardly knows best.

Committed to an outdated value system, D-FENS cannot bridge his nostalgia for the past with the unpleasant realities of the present, and his world comes falling down. Rather than commit suicide, he eventually provokes Prendergast, the detective on his tail, to shoot him dead, so that Beth and Adele might benefit from his life insurance policy. Meanwhile, in the film's final scene, the home video plays on by itself inside the empty house, the master plot to which he was slave, its idealism replaying in an endless loop.

As an existential outsider unable to find a place of his own in an absurd landscape, D-FENS turns to home video as the only experiential gestalt available to stave off utter schizophrenia. As an "orientational metaphor," home video provides an overall system by which to organize physical and cultural experience coherently and purposefully.[42] When the VIT mobilizes these orientational connotations as a semantic unit, it acts as a cognitive schema cueing spectatorial interpretation of the narrative according to the cultural functions and expectations associated with the home mode: community, consensus, and continuity. For example, in *Reality Bites* and *Totally F***ed Up,* films featuring disaffected youth alienated from their biological families, yet in search of families they choose, home videotapes jointly produced by a group of friends, such as Lelaina's documentary and Steven's video diary, immediately evoke expectations of cooperation, togetherness, and purpose; even when these projects fail, their goals remain idealized, if unattainable.

The VIT in the home mode thus provides the personal life trajectories of individual characters with a sense of history, of home, of affiliation. In *Until the End of the World,* Sam frantically crisscrosses the world in search of fragmented video images of friends and relatives in the hope that his mother might order them within the all-encompassing gaze of her new-found vision, like some image custodian organizing "our story," as she puts it, and thus the sense of history that so far had evaded the Farber family. In *Home for the Holidays* (1995), despite and because of his dysfunctional family's inability to get along together for more than five minutes, the father of the household clings to his camcorder, taping their reunions so that later he can review all of his important memories on tape alone in his den, in retrospect transforming his unruly household into a meaningful home. And in *Things to Do in Denver when You're Dead* (1995), "Afterlife Advice," a video service for terminally ill patients to record a series of philosophical speculations accessed later by younger generations seeking wisdom or counsel, offers participants the opportunity to view ties with deceased relatives and thus revive and renew kinship affiliations.

Oedipal Conflict and Resolution: Family Romance and Primal Scene

If the cognitive aspects of family formation are emphasized in these examples where the VIT organizes the connotations of the home mode's social functions, in other films, notably the works of Atom Egoyan, the psychoanalytic aspects of family formation drive the narrative forward. In *Next of Kin* and *Family Viewing,* for example, the VIT organizes the connotations of Freud's Oedipus complex into the dramatic fantasy structures of the family romance and primal scene. Here the VIT functions not merely as a cognitive fulcrum for conscious intention but as a motor for unconscious desire.

At the center of the family romance is Freud's observation during psychoanalysis of a common wish fulfillment narrative among patients who imagined the act of replacing one or both parents with superior figures making up for the perceived shortcomings of their real fathers or mothers. In *Next of Kin,* Egoyan uses video as a medium to channel these desires, its metadiegesis within his filmic narrative the stage on which Peter, the protagonist, acts out his family romance. Dissatisfied with his Waspish family, neither recognized as an adult by his father nor genuinely loved by his mother, Peter seeks greater satisfaction with the Deryans, an Armenian family whose emotional intensity appears to be a welcome substitute. But contrary to Freud's model of fantasy activity, Peter makes the switch literally rather than metaphorically, with the help of a videotaped psychotherapy session he watches by chance before viewing one of the sessions in which he participated with his own parents.

The session attempts to reconcile George Deryan with his daughter Azah, a young adult resisting her father's oppressive dictates. Provoking discourse among the family members, the therapist suggests that Azah represents George's fear that Bedros, their infant son given up for adoption when the Deryans emigrated to the United States, would have rejected his Armenian heritage had he grown to adulthood. Playing Bedros in an exercise, the therapist enrages George by calling him a "lousy foreigner," provoking an attack. Knocked to the floor and pinned down under George's fist, the therapist encourages: "Hit me, George. Hit your son. As hard as you want."

The camera recording this session, initially stable, eventually intrudes on the action in a mobile, handheld style metaphorically illustrating Peter's capacity through the video medium to project his desire directly into the scene. Through the therapist, with whom he identifies (and resembles physically), Peter may vicariously play out both his repressed rage against his father and his family romance of becoming Bedros. Imitating the images that he sees, Peter ultimately embodies them by traveling to Toronto to impersonate Bedros before the Deryan family, who against rational logic willingly embrace him, sutured into their own fantasy by an image cathexis that assuages their guilt for abandoning their real son.

This presentation of Peter's therapy as a process of evaluating images rather than the activities of real life implicitly critiques contemporary epistemology. Egoyan suggests that images are no guarantee of truth, which comes from experience, not from vision. Recognition is therefore always double, both material and mental, a relation video tends to split into arbitrary signifiers and signifieds, such that bodies and identities become exchangeable. In his tape-recorded diary, for instance, Peter realizes that "being alone was easier if you became two people. One part of you would always be the same, like an audience, and the other part would take on different roles, like an actor." That Peter could transmogrify himself as Bedros by casually appropriating the Deryans' tape for his own intentions also suggests the danger of surrendering private experience to video, whose easy access allows unaccountable others to manipulate its content and context.

Playing out his fantasy until the very end, Peter actually leaves his biological family to live with the Deryans. Yet the resolution of his Oedipus complex therefore remains within the realm of fantasy rather than reality, within the Imaginary rather than the Symbolic. Instead of usurping the position of controlling father, Peter merely reaffirms the position of dutiful son, moving laterally into the more comfortable position left open by Bedros. Thus Peter remains subject to Oedipal tensions (illustrated by the "Deryan and Son" T-shirt he wears), but within a family who proffer greater affection in exchange for his subordination. As Peter admits: "It takes more effort to speak what's on your mind than to say, 'Yes, Dad.'"

Passive-aggressive, Peter controls image, not substance, manipulating the Deryans' desire to see their Armenian ethnicity in his own obviously Anglo-Saxon face, a blank screen open to their projections. An embodiment of video's potential for vicarious experience, Peter is a pure simulacrum, even of himself: "For the first time I couldn't tell which side of me was taking which part. It was a bit scary at first. But I'm beginning to like it." Split into signifiers without signifieds, Peter's identity thus capitulates to metonymy rather than achieving the autonomy he had originally sought.

Like the family romance, the primal scene is another Freudian fantasy structure motored by a genealogical regression. In this story, however, subjects imagine overhearing or observing parental sexual intercourse. Although this latent fantasy may never actually be realized outside of unconscious desire, it can manifest itself in symbolic practices. Stanley Milgram, for instance, links the practice of photography to the primal scene's compulsive power to fascinate: "Freudian psychologists might say that the profession of photography is a sublimated form of voyeurism, and underlying every lifelong commitment to photography is some remnant of the desire to catch a glimpse of the primal scene, sexual intercourse with one's parents."[43]

In *Family Viewing,* Egoyan dramatizes similar connections between video and the primal scene. The narrative conflict concerns Van and Stan, son and father, protagonist and antagonist, who compete within the same household to control the archive of videotapes preserving the memories of Van's beloved mother before she fled the family after years of sexual abuse by Stan. In one scene, Van discovers that Stan has been recording sex acts with his live-in lover Sandra over old home video footage of his mother and himself as a young boy. Here the video apparatus functions as the object of a tug-of-war between the younger generation, who needs it to preserve childhood memory, and the older, who employs it to display and extend his power, rewriting personal history in the process. Van would like to rewind his family history, to return to the time when unity with his mother was possible and his father remained outside his primary frame of reference. Thus when he eventually moves out of Stan's apartment, Van switches the tapes with blanks to conserve them and steals his father's VCR, the equivalent of Stan's phallus, in an act of Oedipal usurpation.

Moving into an apartment of his own, Van, like Peter in *Next of Kin*, begins his search for autonomy. And like Peter, Van's first impulse is to resolve his Oedipal conflict in the realm of the Imaginary. Transferring his maternal grandmother, Armen, from a nursing home into his apartment, Van spends much of his time playing back old home videos for her, in particular scenes of her daughter. In this way, Van uses both video and his grandmother as media to channel his desire for unity with his absent mother. Again like Peter, Van confuses control with the manipulation of metonymic images. In one crucial scene, however, he makes the important rite of passage into the Symbolic so that he can finally usurp Stan's position as family patriarch. Discovering among the footage that Stan had not yet erased a grotesque image of his mother bound and gagged as a fetishized sex object, Van realizes for the first time why his mother deserted the family. As though stumbling on a primal scene in which he has unconsciously become father to himself, shocked into action, he takes a more active role from this point onward in the narrative. The video image, therefore, no longer merely a signifier of pre-Oedipal unity, takes on symbolic significance, motivating Van to find his mother and reconstitute the family outside of Stan's frame of reference typically confined to video images.

Before his Oedipal rebellion, Van had been subject entirely to a video-saturated universe: not only were his childhood memories generated by consumer VHS, but the scenes of his contemporary life in his father's apartment were shot in sitcom style with three studio cameras, interspersed with broadcast television and surveillance video footage. Rejecting Stan's world, Van enters into the realm of film, where over the course of his maturation into an independent adult, the video sequences gradually diminish until by the story's end, Van and his mother join together within the same film frame rather than remaining separated by the barrier of video's vicarious illusion of unity. Although a surveillance video camera records the scene, neither do we know the consciousness directing it, nor do we view the action through the video apparatus, suggesting that Van's victory over Stan could occur only from a broader frame of reference. In the film's final image, Stan looks over his shoulder as if caught off guard, his all-powerful position as camcorder operator usurped by Van's working "behind his back," his omniscient frame of reference reframed as just one narrow, subjective perspective.

An important implication of Egoyan's films is that video fails as a medium of interpersonal communication. Despite its appropriation as a tool of psychotherapy in *Next of Kin*, Peter and his parents never bridge their differences in the video modeling sessions. And in *Family Viewing*, video fails to mediate altogether, instead causing irreconcilable conflict between Van and Stan. Thus although video had historically been championed in the sixties and seventies for its potential to connect individuals because of its capacity for simultaneity and feedback, by the eighties and nineties, that hope had been dramatized by cinematic narratives as naive, if not dangerous.

In *Down and Out in Beverly Hills,* yet another Oedipal drama plays out through the video medium, used by the son, Max, as a mode of communication through which to resolve conflicts with Dave, the father. Creating innovative videotapes with the intention to express his fundamental differences from Dave, for whom he prefers to leave cassettes in absentia rather than face him in person, Max believes in the efficacy of video in and of itself to communicate his point of view. Dave, however, of an older generation untrained in the elliptical, fast-paced style of Max's tapes, simply cannot read or interpret them as anything but incoherent, and worse, signs of a disturbed mind. Thus rather than resolve the Oedipal conflict, video becomes the primary apparatus aggravating it: interpersonal communication degenerates to the point where Dave merely shouts at Max to get on with his life while Max hides behind his camcorder, shooting Dave's spectacular tantrums.

One implicit reason for video's inadequacy as a communications medium is its perceived solipsistic nature, often attractive to narcissists engaged more by intercourse with themselves than with others. For example, in *Totally F***ed Up,* Steven proclaims that his video project wants to show "the way things really are," evoking video's ideology of raw experience; nevertheless Steven fails to communicate with much more than his own ego. Each of his interpersonal experiences—of friendship, love, sex— he distances through the viewfinder of his camcorder so that he may objectify and manipulate his associates to his own advantage. In particular, Steven mediates his relationship with his lover Derek to avoid commitment. For instance, after cheating on Derek with another man, Steven confesses his guilt only to himself and his camera. When Derek accidentally discovers the tape and leaves Steven in a rage, rather than meet person to person, Steven responds by mailing Derek a precious, self-serving video letter of apology.

Another narrative explanation for video's failure to communicate exploits its popular perception as a medium for consuming, rather than producing, information. In *Slacker* (1994), a frustrated graduate student, in advance of a thesis defense he knows will end in failure, produces a videotape to protest faculty injustice. At the end of his recording, the student fires a shotgun at the camcorder, perhaps an act of defiance against his committee's surveillance and supervision of his work. In any case, a television freak obsessed with video of every sort dubs and distributes the grad student's tape—not because of its political message but because of its spectacle. Caring little about the student's plight, the "video backpacker" (he wears a monitor on his back) fast-forwards through the spoken editorial portions to relish in the gunfire at the climax. What really matters to him is not the grad student's message or the truth of his situation but that the situation has been mediated. This criterion alone gives the grad student his primary significance.

This sequence from *Slacker* suggests that video access, often rhapsodized in utopian discourse for its democratic potential, may be fraught with perils when the medium falls into the wrong hands. Returning to *Next of Kin*, we see how casually Peter has managed to view a public display of the Deryans' private therapy session; furthermore, although the cassette had been intended by the therapist for a woman named Sarah, the proper recipient of the tape's information, Peter disregards its interpersonal salutation. Of all failures of communication dramatized by the VIT, perhaps this theme—the dangers of unrestricted access—is most common in cinematic narratives. In *American Beauty*, a homophobic father misreads his son's videotape of the neighbor next door as evidence of homoerotic desire, leading to an unnecessarily tragic murder; in *Bad Influence*, when Alex plays a stolen tape of Michael in bed with Claire in front of Michael's fiancée at their engagement dinner, the wedding is called off; in *Totally F***ed Up*, Derek's discovery of Steven's confessional tape results in their estrangement; and in *Sex, Lies, and Videotape*, although Graham convinces Ann that only he has access to the cassettes displaying his subjects' sexual fantasies, John, her husband, breaks into Graham's apartment to watch the tape in which Ann reveals to Graham that she wants a divorce. Thus, in each example, the intention toward interpersonal communication fails as its inverse: permanent separation.

■ ————————————————————————————————

The Autobiographical Impulse: Exposition, Confession, and Analysis

In one respect, video's representation as a solipsistic medium compromises its effectivity as a tool of communication; in another, video's narcissistic connotations empower it as a tool of autobiography. That is, subjects within in cinematic narratives frequently communicate with themselves by using the video apparatus as a catalyst to tell their own stories, confess their private feelings and beliefs, and analyze their images of self. If the VIT fails in its communicative function, it succeeds in its testimonial function.

At a practical level, the VIT's autobiographical impulse provides narrative exposition. Sometimes the exposition is performed in a straightforward manner to advance the plot or develop character, as in *My Life* when Bob tells the story of how he first met his wife Gail, or in *Reality Bites,* where Troy relates the details of his parents' divorce, his father's cancer, and his general philosophy of life. In other cases, the exposition works as a form of shorthand that appropriates aspects of video practice to suggest character motivation in broader strokes. For instance, the opening sequence of *Totally F***ed Up* introduces the cast with snippets of their interviews from Steven's video journal, which later, interspersed throughout the narrative, function like a chorus unifying its fifteen fragments with editorial commentary. Even more suggestive, a home video flashback in *To Die For* succinctly implies the cause of Suzanne Stone's compulsive desire for a television career: already a "natural" in front of the camcorder as a toddler, Suzanne connects her place at the center of her parents' attention with her mediated image on the television monitor, a validation of her self-worth.

As an adult, Suzanne relates the story of her rise to fame as a broadcast journalist by directly addressing the film camera in expository scenes that seem intended for "us," the narrative's audience. Although this technique may seem artificial in contrast to the rest of the film's docudrama realism, by the end of Suzanne's recitation, we discover that she had indeed been directly addressing neither the film camera nor the audience, but her own camcorder and the Hollywood executives she believes want to turn her life story into a movie. Thus Suzanne's exposition is grounded in video practice, which, in keeping with her character, reaffirms the docudrama's verisimilitude.

As a video letter, so to speak, intended for a specific recipient internal to the diegesis, but available to a general audience watching from the outside, Suzanne's exposition functions in a manner similar to epistolary literature. The VIT records for view the internal thoughts that a letter transcribes on the written page. In both cases, exposition is provided not by an implied narrator but by a character who both lives out and then later comments on the action. The self-reflexive nature of the VIT is therefore doubled by the character's retrospective analysis, a metacritique of the narrative diegesis. One of the many ironies of *To Die For,* of course, is that Suzanne's retrospection is entirely unreflexive, so hell-bent on notoriety that she never confesses her guilt as a murderer, let alone comprehends the consequences of her deceit.

In *I've Heard the Mermaids Singing,* however, a narrative also evoking epistolary conventions, Polly, a protagonist as neurotically insecure about her career path as Suzanne is neurotically self-confident, constantly reflects on the pros and cons of every choice she makes in her personal and professional lives. Acting as a catalyst, the video apparatus provokes Polly to confess her daydreams about success and her fears about the future. Like a two-way glass, the video camcorder functions as a screen delivering Polly's confession to an invisible interlocutor, and the video monitor functions as a mirror reflecting Polly back to herself in a confrontation that deepens her self-awareness.

The Big Chill illustrates how video's screen/mirror configuration may split the single subject into both roles as confessor and interlocutor. A drug dealer disenchanted by his present profession, cynical about his prospects for the future, and defensive about his aborted doctoral candidacy in the past, the character of Nick interviews himself with a camcorder, switching back and forth between the positions of interviewer and interviewee. This playacting session allows him distance from himself during the production of the videotape in order to reflect on the chaotic course of his life, and when he views the tape later on a video monitor, even more distance from the position of the interview's audience. We might say that Nick's subjectivity triangulates in this case, offering three perspectives among which he may try to reconstruct his self-identity.

In *Sex, Lies, and Videotape,* Graham and Ann split the roles of confessor and interlocutor between them. At first, Graham takes control behind the camcorder as the interlocutor, provoking Ann to confess that her marriage to John is a sham, that she feels sexually unsatisfied, and that she has fantasized about Graham. Ann admits that years of face-to-face meetings with her psychiatrist have never resolved her problems, implying that the video camera has proved a better medium of therapy. Graham, on the other hand, guilt ridden by his narcissism, which has destroyed his relationships with women in the past, has turned to an obsession with video because he believes the medium to be neutral, distancing, impersonal—a means to buffer women and himself from the (often violent) vicissitudes of his self-serving ego. Yet Ann reminds him of video's interventionist nature; she serves, in fact, as an example of how Graham's tapes have made an important impact on her life. At this point, she grabs the camcorder from Graham, turns its inquisitive eye on him, and inverts their roles as he begins to confess to her the details of his troubled past. Liberated from many of their inhibitions after working through their personal issues on camera, Ann and Graham embrace, enabled to renew their lives together as a couple.

In these films, self-confrontation of one's mediated image develops a deepening awareness of dysfunction that in turn provokes positive behavioral change. In other films, however, the same process can be a threatening, if not shattering, experience leading to anxiety or depression. In *Speaking Parts,* for example, Lisa assists Eddy, a wedding videographer, by taping an interview of the bride. Instructed merely to evoke the bride's sense of excitement, Lisa asks her about the nature of her relationship with the groom, questions that, more appropriate to a counseling session, put the bride in an extremely uncomfortable position. The camcorder's insistent presence, along with Lisa's interrogation, provokes the bride to think more deeply about the groom's love for her, placing a doubt in her mind that the marriage may indeed be a mistake she will regret in the future and finally reducing her to tears. Enraged, the groom accosts Lisa, shoving the camcorder into her face and shouting, "How do you like it? Don't do it to other people!"

Vicarious Experience: Phasing In and Out of Reality

Because Lisa is obsessed by her unrequited love for Lance, she loses her objectivity, channels her own fears about love through the camcorder, and re-projects them onto the bride. Like so many of Egoyan's characters, Lisa uses video as a medium of vicarious experience. Indeed, in the universe of Egoyan's films, "real" life and its mediation cannot readily be detached. In *Family Viewing* especially, until liberated from his father's frame of reference, Van lives out his existence as if he were acting in a time-shifted television program. In one scene, flirting with incestuous desires, Van nearly kisses Sandra on the living-room couch, but before their lips can meet, the image freezes, then rewinds for several seconds, delaying if not preventing the taboo act. This unusual technique, not quite a flashback, suggests the degree to which Van conceives of his life as a series of mediated moments, and his desire to manipulate, to control—even to reverse—the order in which these moments occur as he might the home videotapes that he confuses with his own identity.

Van's obsession with mediating the events of his life is not so different from that of many home video practitioners, except perhaps in magnitude. Who isn't familiar with a friend or relative who enjoys reviewing the past on video more than experiencing the present directly? Who hasn't on one occasion or another directed circumstances to suit the needs of the camera more than those of the participants? Poignantly expressing the motivation driving this desire for mediation, the father in *Home for the Holidays* wishes he could have saved every moment of his life in a videotaped archive to recapture and relive emotions not so easily evoked by his organic memory, dulled by age and soured by the cynicism of bitter experience. Taking up the more literal and consequential effects that an exhaustively recorded life review might entail, *Defending Your Life* (1991) toys with the idea that one's destination after death depends on a defense of every moment in one's past experience, each recorded, logged in, referenced, and queued up for examination.

Even in death, therefore, life may be reincarnated on video. In *My Life,* for example, Bob Jones lives on in his videotapes long after his demise. Using video as a defense mechanism to ward off his fear of death, if not deny it altogether, Bob offers his son not his flesh-and-blood substance as a living father figure, subject to decay and putrescence, but a vicarious substitute electronically embalmed in a timeless cyberspace that reconstitutes the nuclear family and blesses the narrative with a happy ending. Although Bob's afterlife affiliation to his wife and child remains mediated, its parasocial nature psychologically resembles face-to-face interaction. In the film's final scene, when Bob's talking head recorded on video reads a story to his wife and child, they recathect with his image as a thanatic fetish, if you will, transferring the affectivity belonging to his absent body onto its video simulacrum. Bob hasn't died, exactly: what his family seems to mourn instead is the impossibility for increased intimacy and physical contact.

In many cinematic narratives, "video" and "simulacrum" are indeed synonymous. Characters unhappy with the reality of their social situations frequently turn to video to explore fantasies, which when externalized on tape take on an autonomous power that may in some cases be mistaken as authentic experience. For example, in *Jimmy Hollywood,* the eponymous protagonist, who has failed to realize his dream of becoming a legitimate actor, stages an elaborate hoax that finally draws the media attention that has eluded him. Fabricating the character of "Jericho," a street vigilante who takes justice into his own hands, Jimmy plays the role in a series of homemade videotapes that he sends to local news stations, which in turn broadcast his polemical speeches throughout New York City. Drunk with notoriety, Jimmy soon inhabits Jericho's persona, blurring his legitimate identity with an illegitimate substitute, his real and vicarious experiences of self. In *Welcome to the Dollhouse* (1996), Dawn Weiner, a hopelessly awkward adolescent, envies her younger, prettier, and more talented sister, Missy, the pride of her parents. After being humiliated on a home video in which Missy shoves Dawn into a wading pool, earning admiration and laughs from the other family members, an enraged Dawn nearly takes a hammer to Missy's skull later that night in their bedroom. Yet rather than commit murder, Dawn steals the videotape from the VCR and smashes it to bits outside near the garbage, in effect living out her revenge vicariously through Missy's video simulacrum.

The vicarious experience of sex, of course, is the most prevalent simulacrum portrayed by the VIT. Egoyan especially investigates the link between video and sexual fantasy. In *Speaking Parts,* Clara, a screenwriter working through her latent incestuous desire for her dead brother, casts Lance, who resembles him, in a script about his untimely demise. To dramatize Clara's vicarious experience of her brother through Lance, rather than film a conventional sex scene, Egoyan stages a teleconference between the two, who, safely buffered by electronic space, masturbate to orgasm, thus sublimating Clara's desires more deeply. In *Family Viewing,* Stan's sexual gratification depends entirely on mediated fantasies. Like separate audio and video channels that he can manipulate at will, Stan acts out his fantasies between Aline, a disembodied phone actress who supplies the dialogue, and Sandra, a muted live-in lover who supplies the body. Taking her cues from Aline, Sandra acts out Stan's fantasies under the watchful eye of a video camera that Stan may review on a monitor to his satisfaction alone.

As this scene illustrates, the motivating link between sex and video is voyeurism. Yet unlike Metz's notion of cinematic scopophilia, which suppresses the marks of the subject of enunciation whenever possible, the VIT foregrounds rather than disguises the visibility of subject and object. The audience's awareness of the video apparatus undermines their sense of self as authoring agencies, for they witness another's desire as much as their own: when we watch Stan watching Sandra's naked body on video, we watch desire itself as much as its intended object. Although video may provide the distance between subject and object requisite for scopophilia, the object can, as in *Speaking Parts,* return the subject's gaze, blurring the distinction between voyeurism and exhibitionism. And as a third gaze, the audience oscillates between the two, made visible to their own act of looking.

The opening sequence of *Stealing Beauty* (1996) is an odd but interesting miniessay on the voyeuristic nature of cinematic narrative, in particular the story that follows, in which a young woman traveling in Italy blossoms into a fully sexual being desired by nearly every older man she encounters. Liv Tyler, who plays the role of protagonist, is first introduced as a video image framed on all sides by a black border, probably simulating the frame of a video monitor or viewfinder. Many shots record her sleeping on a plane and train voyage to Italy, suggesting the work of a voyeur concerned that his subject remain unaware of his presence. At one point, for instance, when the woman is awake, the camera viewfinder seems to sneak up over her shoulder to see what she is reading but rapidly retracts when she looks back in surprise. In all cases, the identity of the operator behind the camcorder is never visibly revealed, but we can guess he is a voyeur rather than a documentarist or private detective because he focuses his lens primarily on the woman's lips and crotch. Indeed, the sequence even implies that the film's director, Bernardo Bertolucci, may be the consciousness behind the video camera. Eventually discovered by the young woman, who asks him what he's doing, we hear a man with a European accent respond, "This is for you," transferring the tape he has been recording into her possession. Perhaps Bertolucci is implicating his own voyeuristic motivations for shooting an entire narrative around Tyler, who appears in virtually every scene, as well as suggesting that video, like his film itself, is at once stealing Tyler's beauty from her as a media artifact outside her control for the purpose of satisfying older men's masturbatory fantasies and returning her beauty as a gift, preserved eternally against the ravages of time.

Although ambiguous, the opening scene of *Stealing Beauty* may serve as little more than an alibi inviting rather than critiquing a voyeuristic response to the film. Therefore the VIT's self-reflexive potential to expose voyeurism may be recuperated in its service. For example, although *Henry: Portrait of a Serial Killer* portrays its protagonists as psychotic hoodlums, the ambivalent nature of the videotaped sequences of their disturbing behavior may be either rejected as morally repulsive or enjoyed as a vicarious thrill. In one particularly outrageous sequence, Henry and Otis break into a suburban home, kill the inhabitants, and videotape the entire atrocity. The degree to which we identify with Henry, who operates the camcorder, will determine the degree to which we condemn his actions—fair enough. But at one point in the sequence, when Henry puts the camera down on the floor to dispatch a young boy who walks into the house, the subject position behind the lens is left wide open, leaving us at least three possible responses: (1) we continue to suture the lens into Henry's consciousness, which is momentarily in absentia; (2) we conceive of the camera as a neutral recording device, shooting the action in an impersonal documentary mode that we need not identify with our own intentions; or (3) we actively take up the lens with our own consciousness and must negotiate our feelings about voluntarily observing this unspeakable scene of violence. Just at the point at which we may be forced to acknowledge our own fascination by the murders, Henry picks the camera up and reprimands Otis for sexually molesting a woman's corpse, just in time inflecting a truly ambivalent subject position with moral outrage, and distancing us, the audience, from any possible sense that we too have been accessories to the crime.

An implicit theme of *Henry*, linked to its theme of voyeurism and dramatized by its socially retarded protagonists, is that a dependency on video as a medium of vicarious experience is symptomatic of an unhealthy relation to the real world, or at the very least an immature phase of development that must be surpassed. The examples proliferate: in *Until the End of the World*, Claire must give up reviewing her dreams on video and read books instead before she can take responsibility for herself and the world; in *Family Viewing*, Van must reframe himself outside of Stan's video surveillance to usurp the father's position and take control of his own life as an adult; in *Down and Out in Beverly Hills*, Max must dispense with his video essays and communicate directly to Dave to reconcile their conflicts face-to-face; in *Shooting Lily*, the husband must promise his wife that he will stop recording her incessantly to repair their estranged marriage; in *Sliver*, Carly saves Zeke from his self-imprisoning solipsism by blowing up his video surveillance room and directing him to "get a life"; and in *Sex, Lies, and Videotape*, Graham must smash his TV monitor, camcorder, and videocassettes to bits—in effect go cold turkey—so that he can beat his video addiction. In each case, and there are many others, a proper narrative resolution may transpire only under the omnipotent eye of the film camera's omniscient lens. Video, a vicarious and precarious alternative, is a siren whose seductions must be navigated at one's own peril, at least in the myths of cinema.

Notes

Introduction

1. Rudolf Arnheim, *Film as Art* (Berkeley: University of California Press, 1957), 1.
2. André Bazin, "The Myth of Total Cinema," in *What Is Cinema?* vol. 1, ed. and trans. Hugh Gray (Berkeley: University of California Press, 1967), 17–22.
3. Fredric Jameson, "Surrealism without the Unconscious," in *Postmodernism, or The Cultural Logic of Late Capitalism* (Durham: Duke University Press, 1991), 71.
4. Randal Johnson, "Editor's Introduction: Pierre Bourdieu on Art, Literature, and Culture," in *The Field of Cultural Production: Essays on Art and Literature*, by Pierre Bourdieu, ed. Randal Johnson (New York: Columbia University Press, 1993), 6.
5. Johnson, 16.
6. Gilles Deleuze and Félix Guattari, *Anti-Oedipus: Capitalism and Schizophrenia*, trans. Robert Hurley, Mark Seem, and Helen R. Lane (Minneapolis: University of Minnesota Press, 1983), 117.
7. Stuart Hall, "Notes on Deconstructing 'the Popular,'" in *People's History and Socialist Theory*, ed. Raphael Samuel (London: Routledge, 1981), 235.
8. Johnson, 4.
9. Jean-Louis Comolli, "Technique and Ideology: Camera, Perspective, Depth of Field (Parts 3 and 4)," in *Narrative, Apparatus, Ideology: A Film Theory Reader*, ed. Philip Rosen (New York: Columbia University Press, 1986), 424.

1. What Is Video?

1. Marita Sturken, "Paradox in the Evolution of an Art Form: Great Expectations and the Making of a History," in *Illuminating Video: An Essential Guide to Video Art*, ed. Doug Hall and Sally Jo Fifer (New York: Aperture, 1990), 118.
2. Stuart Marshall, "Television/Video: Technology/Forms," *Afterimage* 8–9 (spring 1981): 71–85.
3. Marshall McLuhan, *Understanding Media: The Extensions of Man* (New York: New American Library, 1964).
4. For a more detailed distinction between mechanistic and symptomatic causality, as well as their critique, see Jennifer Daryl Slack, *Communication Technologies and Society: Conceptions of Causality and the Politics of Technological Intervention* (Norwood, N.J.: Ablex, 1984), 52–62.
5. For a more detailed account of Williams's concept of "mobile privatization," see "The Technology and the Society," chapter 1 of *Television: Technology and Cultural Form* (Hanover, N.H.: University Press of New England, 1974), 3–25.
6. Lynn Spigel, introduction to *Television: Technology and Cultural Form*, by Raymond Williams (Hanover, N.H.: University Press of New England, 1992), xvii.
7. Noël Carroll, "Medium Specificity Arguments and Self-Consciously Invented Arts: Film, Video, and Photography," *Millennium Film Journal* 14–15 (fall–winter 1984–1985): 127–53.
8. Roy Armes, *On Video* (London: Routledge, 1988), 1.
9. For this succinct summation of these strategies, I am indebted to David E. James, "inTerVention: The Contexts of Negation for Video and Video-Criticism," *Millennium Film Journal* 20–21 (fall–winter 1988–1989): 53.

10. For a more detailed historical account of activist video, see Deirdre Boyle, *Subject to Change: Guerrilla Television Revisited* (New York: Oxford University Press, 1997).

11. Hans Magnus Enzensberger, "Constituents of a Theory of Media," in *Video Culture: A Critical Investigation*, ed. John G. Hanhardt (New York: Visual Studies Workshop Press, 1986), 97–99.

12. John G. Hanhardt, introduction to *Video Culture: A Critical Investigation*, ed. John G. Hanhardt (New York: Visual Studies Workshop Press, 1986), 16.

13. See Stuart Marshall, "Video: From Art to Independence: A Short History of a New Technology," *Screen* 26, no. 2 (March–April 1985): 66–71.

14. Bruce Kurtz, "Video Is Being Invented," *Arts Magazine* 47 (December–January 1973): 38.

15. Gregory Battcock, introduction to *New Artists Video: A Critical Anthology*, ed. Gregory Battcock (New York: E. P. Dutton, 1978), xviii.

16. Sean Cubitt, *Videography: Video Media as Art and Culture* (New York: St. Martin's Press, 1993), 33.

17. John Belton, "Looking through Video: The Psychology of Video and Film," in *Resolutions: Contemporary Video Practices*, ed. Michael Renov and Erika Suderburg (Minneapolis: University of Minnesota Press, 1996), 71 n. 1.

18. Belton, 66.

19. Rosalind Krauss, "Video: The Aesthetics of Narcissism," in *Video Culture: A Critical Investigation*, ed. John G. Hanhardt (New York: Visual Studies Workshop Press, 1986), 180.

20. Krauss, 181.

21. Toward the end of her article, Krauss does include a discussion of works that "run counter" to her argument, yet she tends to read them as self-reflexive negations of video art's fundamental narcissism, specifically from which they attempt to find critical distance.

22. Michael Renov, "The Subject in History: The New Autobiography in Film and Video," *Afterimage* (summer 1989): 5.

23. Michael Renov, "Video Confessions," in *Resolutions: Contemporary Video Practices*, ed. Michael Renov and Erika Suderburg (Minneapolis: University of Minnesota Press, 1996), 78–101.

24. Ernest Larsen, "For an Impure Cinevideo," *Independent* 13, no. 4 (May 1990): 26.

25. Frank Rickett, "Multimedia," in *Future Visions: New Technologies of the Screen*, ed. Philip Hayward and Tana Wollen (London: BFI, 1993), 74.

26. Lili Berko, "Video: In Search of a Discourse," *Quarterly Review of Film Studies* 10, no. 4 (1989): 289–307.

27. Maureen Turim, "The Cultural Logic of Video," in *Illuminating Video: An Essential Guide to Video Art*, ed. Doug Hall and Sally Jo Fifer (New York: Aperture, 1990), 332.

28. Fredric Jameson, "Surrealism without the Unconscious," in *Postmodernism, or The Cultural Logic of Late Capitalism* (Durham: Duke University Press, 1991), 71.

29. For a more detailed elaboration of this critique of Jameson, see Lawrence Grossberg, "Putting the Pop Back into Postmodernism," in *Universal Abandon? The Politics of Postmodernism*, ed. Andrew Ross (Minneapolis: University of Minnesota Press, 1988), 173–74.

30. See Sean Cubitt, *Timeshift: On Video Culture* (New York: Routledge, 1991), 122–23.

31. For an interesting elaboration of Jameson's disbelief in the cultural efficacy of new or unfamiliar practices, see Nicholas Zurbrugg, "Jameson's Complaint: Video-Art and the Intertextual 'Time-Wall,'" *Screen* 32, no. 1 (spring 1991): 32.

32. Steven Best and Douglas Kellner, *Postmodern Theory: Critical Interrogations* (New York: Guilford Press, 1991), 121.

33. Andrew Dewdney and Frank Boyd, "Television, Computers, Technology, and Cultural Form," in *The Photographic Image in Digital Culture*, ed. Martin Lister (New York: Routledge, 1995), 151.

34. Raymond Williams, *Marxism and Literature* (New York: Oxford University Press, 1977), 120–27.

35. "Structures exhibit tendencies—lines of force, openings and closures which constrain, shape, channel and in that sense, 'determine.' But they cannot determine in the harder sense of fix absolutely, guarantee." Stuart Hall, "Signification, Representation, Ideology: Althusser and the Post-structuralist Debates," *Critical Studies in Mass Communication* 2, no. 2 (1985): 96.

36. See, in particular, Slack, 81–92.

37. Slack, 90.

38. Mark Schubin, "Film vs. Tape," *Videography* 10, no. 4 (April 1985): 30.

39. Quoted in Nicholas Garnham, *Capitalism and Communication: Global Culture and the Economics of Information* (London: Sage, 1990), 6.

40. Michel de Certeau, *The Practice of Everyday Life*, trans. Steven Rendall (Berkeley: University of California Press, 1984), 32.

41. Williams, *Television*, 13.

42. André Bazin, "The Myth of Total Cinema," in *What Is Cinema?* vol. 1, ed. and trans. Hugh Gray (Berkeley: University of California Press, 1967), 17.

43. Jean-Louis Baudry, "The Apparatus: Metapsychological Approaches to the Impression of Reality

in Cinema," in *Film Theory and Criticism: Introductory Readings*, ed. Gerald Mast, Marshall Cohen, and Leo Braudy, 4th ed. (New York: Oxford University Press, 1992), 705.

44. Stephen Heath, "The Cinematic Apparatus: Technology as Historical and Cultural Form," in *The Cinematic Apparatus*, ed. Teresa de Lauretis and Stephen Heath (New York: St. Martin's Press, 1980), 1–13.

45. William Boddy, "The Studios Move into Prime Time: Hollywood and the Television Industry in the 1950s," *Cinema Journal* 24, no. 4 (summer 1985): 29.

46. Jane Feuer, "The Concept of Live Television: Ontology as Ideology," in *Regarding Television: Critical Approaches—an Anthology*, ed. E. Ann Kaplan (Los Angeles: American Film Institute, 1983), 12–22.

47. George Lakoff and Mark Johnson, *Metaphors We Live By* (Chicago: University of Chicago Press, 1980), 146.

48. Robert C. Allen, "Introduction to the Second Edition: More Talk about TV," in *Channels of Discourse, Reassembled: Television and Contemporary Criticism*, ed. Robert C. Allen (Chapel Hill: University of North Carolina Press, 1992), 3.

49. Chris Jenks, "The Centrality of the Eye in Western Culture: An Introduction," in *Visual Culture*, ed. Chris Jenks (New York: Routledge, 1995), 8–9.

50. Nelson Goodman, *Ways of Worldmaking* (Indianapolis: Hackett, 1978), 7–8.

51. George Lakoff, *Women, Fire, and Dangerous Things: What Categories Reveal about the Mind* (Chicago: University of Chicago Press, 1987), 38.

52. Cubitt, *Timeshift*, 106.

53. Harald A. Stadler, "Film as Experience: Phenomenological Concepts in Cinema and Television Studies," *Quarterly Review of Film and Video* 12, no. 3 (1990): 39.

54. Victor Turner, *Dramas, Fields, and Metaphors: Symbolic Action in Human Society* (Ithaca: Cornell University Press, 1974), 30–31.

55. Renee Hobbs, "Television and the Shaping of Cognitive Skills," in *Video Icons and Values*, ed. Alan M. Olson, Christopher Parr, and Debra Parr (Albany: State University of New York Press, 1991), 34.

56. Gilles Deleuze, *Cinema 1: The Movement-Image* and *Cinema 2: The Time-Image* (Minneapolis: University of Minnesota Press, 1991).

57. Pierre Bourdieu, *Distinction: A Social Critique of the Judgement of Taste*, trans. Richard Nice (Cambridge: Harvard University Press, 1984), 467.

58. Sturken, 112–13.

59. Bourdieu, *Distinction*, 478.

60. Raymond Williams, "Realism, Naturalism, and Their Alternatives," *Cine-Tracts* 1, no. 3 (1977–1978): 4.

2. From Reel Families to Families We Choose

1. See in particular the *Journal of Film and Video* 38, nos. 3–4 (summer–fall 1986), which self-consciously inaugurated and advocated the introduction of home mode discourse into cinema studies.

2. Eugene Marlow and Eugene Secunda, *Shifting Time and Space: The Story of Videotape* (New York: Praeger, 1991), 147.

3. Randall Tierney, "Pieces of 8," *American Film* 15, no. 15 (December 1990): 50.

4. Patricia R. Zimmermann, *Reel Families: A Social History of Amateur Film* (Bloomington: Indiana University Press, 1995), 150.

5. Richard Chalfen, *Snapshot Versions of Life* (Bowling Green: Bowling Green State University Popular Press, 1987).

6. Chalfen, 6.

7. Chalfen does acknowledge that his study "has developed from a relatively static synchronic look at pictorial forms" and recommends that "much work remains to be done on how conventions, sanctions, and related behaviors change *diachronically*—through time—and how technical, social, and cultural factors contribute to such changes" (97).

8. Chalfen, 165–66.

9. Quoted in Chalfen, 33.

10. For purposes of clarity and simplicity, I am comparing home movie technology to analog video technology, which was the predominant format at the time Chalfen published *Snapshot Versions of Life*. More significantly, digital video itself can be contrasted to analog video in a similar fashion, since the newer technology may have the capacity to reorient home mode practice, such as increasing the incentive to edit footage using a PC and affordable software. By keeping my comparison of film and video to the substrates of celluloid and electromagnetic tape, I foreground their significant differences in use value despite sharing analog formats.

11. See Chalfen, 93–99.
12. Zimmermann, 133, 122.
13. Mark Poster, *Critical Theory of the Family* (New York: Seabury Press, 1978), 164.
14. George Masnick and Mary Jo Bane, *The Nation's Families: 1960–1990* (Boston: Auburn House, 1980), 98.
15. Arlene Skolnick, *Embattled Paradise: The American Family in an Age of Uncertainty* (New York: HarperCollins, 1991), 51–52.
16. Steven Mintz and Susan Kellogg, *Domestic Revolutions: A Social History of American Family Life* (New York: Free Press, 1988), 178.
17. Skolnick, 129.
18. Kath Weston, "The Politics of Gay Families," in *Rethinking the Family: Some Feminist Questions*, rev. ed., ed. Barrie Thorne and Marilyn Yalom (Boston: Northeastern University Press, 1992), 137.
19. Skolnick, 189.
20. Chris Jenks, *Culture* (New York: Routledge, 1993), 116.
21. See Chalfen, especially 4–16.
22. For a concise summary of the interest theory of ideology, see Clifford Geertz, *The Interpretation of Cultures* (New York: HarperCollins, 1973), 201–3.
23. In his article on "photoconsumerism," Barry King contends that advertising seeks to assimilate all social uses of amateur photography to its own norms and objectives, and that an intention toward domestic representation indicates only our own duplicity: "However much the maps we provide seem configured on the terrain of our own intimacy, it is a visualization rendered according to the geometry of advertising." See "Photo-consumerism and Mnemonic Labor: Capturing the 'Kodak Moment,'" *Afterimage* 21, no. 2 (September 1993): 13. Don Slater confirms that the "impoverishment" of amateur photography "has actually been carried out in large part precisely through the high-pressure mass marketing of photographic equipment. It has been restricted in the course of its very proliferation, through its technical form, through its retailing, through the 'training' of consumers through advertising, the photo press and other publicity organs of the photographic manufacturing industry." See "Marketing Mass Photography," in *Language, Image, Media*, ed. Howard Davis and Paul Walton (New York: St. Martin's Press, 1983), 247.
24. "The public discourse on amateur film functions as a form of social control, because it harnesses subjectivity, imagination, and spontaneity within the more privatized contexts of leisure and family life." Zimmermann, 4.
25. Laurie Ouellette, "Camcorders R Us," *Independent* 17, no. 4 (May 1994): 34.
26. Ouellette, 37.
27. Laurie Ouellette, "Camcorder Dos and Don'ts: Popular Discourses on Amateur Video and Participatory Television," *Velvet Light Trap* 36 (fall 1995): 34.
28. Zimmermann, xiii.
29. Chalfen, 54.
30. Pierre Bourdieu, *Photography: A Middle-Brow Art*, trans. Shaun Whiteside (Stanford: Stanford University Press, 1990), 7.
31. Don Slater provides a concise summary of this position: "The history and dynamics of the photographic industry and that of the family at leisure have to be analysed together, as part of the same process: the social relations which constitute families (and which seem so naturally amenable to certain uses of photography) were themselves determined by the same forces which determined photographic production and marketing in the first place. The specific form of photography inserted into the family is only one instance of an evolving consumerist system colonizing the very structure which was created by the forces which created it—the overall relations of consumption necessitated by the relations of capitalist production" (256).
32. Raymond Williams, *Marxism and Literature* (New York: Oxford University Press, 1977), 106.
33. Christine Tamblyn, "Qualifying the Quotidian: Artist's Video and the Production of Social Space," in *Resolutions: Contemporary Video Practices*, ed. Michael Renov and Erika Suderburg (Minneapolis: University of Minnesota Press, 1996), 13.
34. "The Great *Video Review* Shootoff," *Video Review* 12, no. 6 (September 1991): 18.
35. "Carry a Camcorder," *Video Maker* 2 (November 1989): 24.
36. Pierre Bourdieu, *Outline of a Theory of Practice*, trans. Richard Nice (Cambridge: Cambridge University Press, 1977).
37. Bourdieu, *Photography*, 20.
38. Bourdieu, *Outline*, 79.
39. Bourdieu tends to homogenize diversity within classes, a flaw that has been criticized but can be corrected with little damage to the overall value of his theory. See, for example, John Frow, "Accounting for Tastes: Some Problems in Bourdieu's Sociology of Culture," *Cultural Studies* 1, no. 1 (January 1987): 59–73.
40. Bourdieu, *Outline*, 80.

41. For a summary of strain theory, see Geertz, 203–7.
42. See, for instance, Judith Williamson's argument in "Family, Education, Photography," in *Consuming Passions: The Dynamics of Popular Culture* (New York: Marion Boyars, 1986), 115–26.
43. Roger Silverstone posits the issue as a question: "'Did you create that or did you find it?' This seems to me to be the key question at the centre of the problematic of everyday life. . . . If it were to be answered—and popular culture does answer in its own way—then the reply would be both and neither. Everyday culture is in this sense, and within this paradox, transitional. . . . And the terms of that paradox—the found object and the created object—the imposed meaning and the selected meanings—the controlled behaviour and the free—the meaningless and the meaningful—the passive and the active—are in constant tension." *Television and Everyday Life* (New York: Routledge, 1994), 164.
44. Roger Silverstone, Eric Hirsch, and David Morley, "Information and Communication Technologies and the Moral Economy of the Household," in *Consuming Technologies: Media and Information in Domestic Spaces,* ed. Roger Silverstone and Eric Hirsch (New York: Routledge, 1992), 16. For a more detailed discussion of the process of conversion, see 21–26.
45. Advancing the notion of the biography of commodities, Igor Kopytoff has proposed a processional mode of commoditization, in which objects may be moved both into and out of the commodity state—that a commodity is not one kind of thing rather than another but one phase in the life of some things. See Arjun Appadurai, "Introduction: Commodities and the Politics of Value," in *The Social Life of Things: Commodities in Cultural Perspective,* ed. Arjun Appadurai (Cambridge: Cambridge University Press, 1986), 17.
46. See Chalfen, 50.
47. Malcolm Chase and Christopher Shaw, "The Dimensions of Nostalgia," in *The Imagined Past: History and Nostalgia,* ed. Christopher Shaw and Malcolm Chase (New York: Manchester University Press, 1989), 11.
48. Mihaly Csikszentmihalyi and Eugene Rochberg-Halton, *The Meaning of Things: Domestic Symbols and the Self* (New York: Cambridge University Press, 1981), 119.
49. See Silverstone, 28, for his summary of these domains borrowed from the work of Judith and Andrew Sixsmith.
50. Silverstone, 45.
51. Barbara Myerhoff, "Rites of Passage: Process and Paradox," in *Celebration: Studies in Festivity and Ritual,* ed. Victor Turner (Washington, D.C.: Smithsonian Institute Press, 1982), 109.
52. Sol Worth, *Studying Visual Communication,* ed. Larry Gross (Philadelphia: University of Pennsylvania Press, 1981), 195.

3. Modes of Distinction

1. "Perhaps no video field forged by new technology has ever been as potentially large as that of professional special-event videography. Powered by the arrival of such relatively low-cost, high-performance equipment . . . this field is creating opportunities to address the special video needs of a world in which everyone has a VCR." "The Professional Special-Event Videographer's Handbook," *Videography* 16, no. 1 (January 1991): 1.
2. Paul D. Kennamer, founder of the Society of Professional Videographers in Huntsville, Alabama, estimated in 1991 that over 100,000 videographers derived financial gain from event videography. See "The Professional Special-Event Videographer's Handbook," 1.
3. Michael Renov, "Toward a Poetics of Documentary," in *Theorizing Documentary,* ed. Michael Renov (New York: Routledge, 1993), 22–35.
4. Vivian Sobchack, "Inscribing Ethical Space: Ten Propositions on Death, Representation, and Documentary," *Quarterly Review of Film Studies* 9, no. 4 (1984): 293.
5. Stephen Neale, *Genre* (London: BFI, 1987), 10.
6. Sol Worth, *Studying Visual Communication,* ed. Larry Gross (Philadelphia: University of Pennsylvania Press, 1981), 138.
7. Pierre Bourdieu, *Photography: A Middle-Brow Art,* trans. Shaun Whiteside (Stanford: Stanford University Press, 1990), 38–39.
8. Bourdieu, *Photography,* 33.
9. Arthur Applebee, *The Child's Concept of Story: Ages Two to Seventeen* (Chicago: University of Chicago Press, 1978), 20–24.
10. Bourdieu, *Photography,* 8.
11. The following summary of Jean-Pierre Meunier's phenomenology of the *film-souvenir* has been condensed from Vivian Sobchack, "Toward a Phenomenology of Nonfictional Film Experience," in *Collecting Visible Evidence,* ed. Jane M. Gaines and Michael Renov (Minneapolis: University of Minnesota Press, 1999), 241–54.

12. Pierre Bourdieu, *The Field of Cultural Production: Essays on Art and Literature*, ed. Randal Johnson (New York: Columbia University Press, 1993), 67.

13. See Maya Deren, "Amateur vs. Professional," *Film Culture* 39 (1965): 45–46; and "Cinematography: The Creative Use of Reality," in *Film Theory and Criticism: Introductory Readings*, ed. Gerald Mast, Marshall Cohen, and Leo Braudy, 4th ed. (New York: Oxford University Press, 1992), 59–70.

14. "Bridging the aesthetic and the existential, film became identified with his life and coextensive with it, simultaneously his vocation and avocation, his work and play." David E. James, *Allegories of Cinema: American Film in the Sixties* (Princeton, N.J.: Princeton University Press, 1989), 37.

15. "I believe any art of the cinema must inevitably arise from the amateur, 'home-movie' making medium." Stan Brakhage, "In Defense of the 'Amateur' Filmmaker," *Filmmakers Newsletter* (summer 1971): 24.

16. James, 47.

17. Brakhage, 25.

18. For a more detailed elaboration of the mnemonic function of Brakhage's films, see P. Adams Sitney, "Autobiography in Avant-Garde Film," in *The Avant-Garde Film: A Reader of Theory and Criticism*, ed. P. Adams Sitney (New York: New York University Press, 1978), 208–28.

19. Jeffrey K. Ruoff, "Home Movies of the Avant-Garde: Jonas Mekas and the New York Art World," in *To Free the Cinema: Jonas Mekas and the New York Underground*, ed. David E. James (Princeton, N.J.: Princeton University Press, 1992), 294–311.

20. Peter Bürger, *Theory of the Avant-Garde*, trans. Michael Shaw (Minneapolis: University of Minnesota Press, 1984), 50.

21. See Victor Turner, *From Ritual to Theatre: The Human Seriousness of Play* (New York: Performing Arts Journal Publications, 1982), 36–40.

22. "Leisure can be conceived of as a betwixt-and-between, a neither-this-nor-that domain between two spells of work or between occupational and familial and civic activity." Turner, 40.

23. "To explain the public's attraction to a medium, one must look not only for ideological manipulation but also for the kernel of utopian fantasy whereby the medium constitutes itself as a projected fulfillment of what is desired and absent within the status quo." Robert Stam, *Subversive Pleasures: Bakhtin, Cultural Criticism, and Film* (Baltimore: Johns Hopkins University Press, 1989), 224.

24. Dziga Vertov, "The Writings of Dziga Vertov," trans. S. Brody, in *Film Culture Reader*, ed. P. Adams Sitney (New York: Praeger, 1970), 359.

25. Jay Ruby, "Speaking For, Speaking About, Speaking With, or Speaking Alongside: An Anthropological and Documentary Dilemma," *Journal of Film and Video* 44, nos. 1–2 (spring–summer 1992): 55.

26. Roger Odin, "For a Semio-pragmatics of Film," in *The Film Spectator: From Sign to Mind*, ed. Warren Buckland (Amsterdam: Amsterdam University Press, 1995), 217.

27. Andreas Huyssen, *After the Great Divide: Modernism, Mass Culture, Postmodernism* (Bloomington: Indiana University Press, 1986), 13–14.

28. See Hans Magnus Enzensberger, "Constituents of a Theory of Media," in *Video Culture: A Critical Investigation*, ed. John G. Hanhardt (New York: Visual Studies Workshop Press, 1986), 96–123.

29. Writes Enzensberger: "Anyone who expects to be emancipated by technological hardware, or by a system of hardware however structured, is the victim of an obscure belief in progress" (107).

30. Nicholas Garnham, "The Myths of Video: A Disciplinary Reminder," in *Capitalism and Communication: Global Culture and the Economics of Information*, ed. Fred Inglis (London: Sage Publications, 1990), 64–69.

31. Jennifer Daryl Slack, *Communication Technologies and Society: Conceptions of Causality and the Politics of Technological Intervention* (Norwood, N.J.: Ablex, 1984), 30–39.

32. See Slack, 141–47, for her critique of the goals of communications revolutions.

33. A note on research: As a former wedding videographer, I first formulated intuitive hypotheses about event videography in the mid-1980s when working to understand how, why, and for whom customers wanted their ceremonies preserved. Since then, innovations in technology and marketing have required that I substantiate these early conjectures with primary research and interviews with current practitioners. An invaluable source of information, the Wedding Video Expo held 26–29 July 1993 in Las Vegas, Nevada, offered seminars, panel discussions, trade shows, and competitions for those in attendance. Many of my theoretical speculations regarding event video production methods and ideologies have been grounded in the statements and practices of the industry's leading figures, who have led the exposition's series of presentations, demonstrated examples of their work, and written articles in various trade publications.

34. My summaries of professionalism's characteristics are indebted to Geoff Esland, "Professions and Professionalism," in *The Politics of Work and Occupations*, ed. Geoff Esland and Graeme Salaman (Toronto: University of Toronto Press, 1980), 213–50; and Magali Sarfatti Larson, *The Rise of Professionalism: A Sociological Analysis* (Berkeley: University of California Press, 1977).

35. "Out-of-home video services are such a new field that anyone with a VCR and a camera can claim to be a videographer, the way anyone with a typewriter can claim to be a writer. No exams have to be passed, and no initials are needed after your name." Frank Lovece, "Home Shooting: Make Money with Your Camera," *Video* 8, no. 12 (March 1985): 83.
36. Event videographer Danny O'Keefe, of Dancel Productions, New Orleans, quoted in "The Professional Special-Event Videographer's Handbook," 3.
37. I adopt the masculine pronoun here not in deference to phallocentric conventions of English prose but in reference to the event videography industry's predominantly patriarchal character. Currently the ratio of men to women in local video associations nationwide is approximately 9:1. Thus although the wedding video is marketed as a "woman's genre," its producers are primarily men.
38. "Most professions produce intangible goods: their product, in other words, is only formally alienable and is inextricably bound to the person and the personality of the producer. It follows, therefore, that *the producers themselves have to be produced* if their products or commodities are to be given a distinctive form." Larson, 14.
39. See Michel de Certeau, *The Practice of Everyday Life*, trans. Steven Rendall (Berkeley: University of California Press, 1984), 6–8.
40. See Esland, 239–50.
41. Bourdieu, *Photography*, 47.
42. Pierre Bourdieu, *Distinction: A Social Critique of the Judgement of Taste*, trans. Richard Nice (Cambridge: Harvard University Press, 1984), 60.
43. Thorstein Veblen, *The Theory of the Leisure Class* (New York: Dover, 1994), 94.
44. "One of the reasons for the lack of respect in the past is that wedding videographers have not earned similar income levels relative to other video producers. Wedding video is extremely underpriced in most areas of the country." Roy Chapman, quoted in "They Shoot Weddings, Don't They?" *Videography* 17, no. 1 (January 1992): 72.
45. See Bourdieu, *Distinction*, 147–54.
46. Randal Johnson, "Pierre Bourdieu on Art, Literature, and Culture," in *The Field of Cultural Production: Essays on Art and Literature*, by Pierre Bourdieu, ed. Randal Johnson (New York: Columbia University Press, 1993), 7.
47. "The Professional Special-Event Videographer's Handbook," 13.
48. Quoted in "The Professional Special-Event Videographer's Handbook," 3.
49. Patricia R. Zimmermann, *Reel Families: A Social History of Amateur Film* (Bloomington: Indiana University Press, 1995), 118.
50. Quoted in "The Professional Special-Event Videographer's Handbook," 3.
51. Zimmermann, *Reel*, 118.
52. "The Professional Special-Event Videographer's Handbook," 4–5.
53. Esland, 245.
54. "Uncle Charlie will always be around. The real question is will the amateurs prevent the professionals from succeeding financially in this business? This [is] a marketing problem that can be overcome." Roy Chapman, quoted in "They Shoot," 72.
55. James, *Allegories*, 5.
56. James, *Allegories*, 14.
57. For a more detailed elaboration of his concept of allegories of cinema, see James, *Allegories*, 12–14.
58. DeeDee Halleck, "Watch Out, Dick Tracy! Popular Video in the Wake of the *Exxon Valdez*," in *Technoculture*, ed. Constance Penley and Andrew Ross (Minneapolis: University of Minnesota Press, 1991), 219.
59. Michele Barrett, "The Place of Aesthetics in Marxist Culture," in *Marxism and the Interpretation of Culture*, ed. Cary Nelson and Lawrence Grossberg (Chicago: University of Illinois Press, 1988), 702.
60. See Nicholas Garnham, *Capitalism and Communication: Global Culture and the Economics of Information* (London: Sage, 1990), 43–44.
61. "When consulted, local professionals give the amateur back the image of photography that he already has, confirming it with their authority." Bourdieu, *Photography*, 172.
62. Richard Chalfen, *Snapshot Versions of Life* (Bowling Green: Bowling Green State University Popular Press, 1987), 60.
63. Ruby, 51.
64. The majority of seminars at the Wedding Video Expo focused on marketing strategies, for example: "How to Create a Demand for Your Services," "Increase Bookings with Your Own Bridal Information Video,'" "Sensational Selling," "Direct Marketing," and "Maximizing Your Profits."
65. Kathy Charmaz, *The Social Reality of Death: Death in Contemporary America* (Reading, Mass.: Addison-Wesley, 1980), 188.

66. Charmaz, 191.
67. See Robert W. Habenstein, "Sociology of Occupations: The Case of the American Funeral Director," in *Human Behavior and Social Processes: An Interactionist Approach*, ed. Arnold M. Rose (Boston: Houghton Mifflin, 1962), 225–46.
68. Described by Jay Ruby in *Secure the Shadow: Death and Photography in America* (Cambridge: MIT Press, 1995), 134.
69. Joan Crawford, interview by author, 14 March 1996.
70. Ruby, *Secure*, 157.
71. This analysis has been informed generally by Richard G. Dumont and Dennis C. Foss, *The American View of Death: Acceptance or Denial?* (Cambridge, Mass.: Schenkman, 1972); and Philippe Aries, *Western Attitudes toward Death: From the Middle Ages to the Present*, trans. Patricia M. Ranum (Baltimore: Johns Hopkins University Press, 1974).
72. See Vivian Sobchack, "Inscribing Ethical Space: Ten Propositions on Death, Representation, and Documentary," *Quarterly Review of Film Studies* 9, no. 4 (1984): 287–88.
73. Peter Metcalf and Richard Huntington, *Celebrations of Death: The Anthropology of Mortuary Ritual*, 2d ed. (Cambridge: Cambridge University Press, 1991), 201.
74. Ruby, *Secure*, 1.
75. André Bazin, "The Ontology of the Photographic Image," in *What Is Cinema?* vol. 1, trans. and ed. Hugh Gray (Berkeley: University of California Press, 1967), 9.
76. Garrett Stewart, "Photo-gravure: Death, Photography, and Film Narrative," *Wide Angle* 9, no. 1 (1987): 12.
77. Crawford interview.
78. Crawford interview.

4. Family Resemblances

1. Ludwig Wittgenstein, *The Blue and Brown Books* (New York: Harper and Row, 1958), 17.
2. As a mechanism of set theory, the recognition of family resemblances has been cited by psychologists, anthropologists, and sociologists as an archetypal foundation for constructing categories. Jean Piaget, for example, has suggested that the development of cognitive operations such as serial ordering and organizational grouping is guided by the child's logical understanding of actual family dynamics. For a concise discussion of Piaget's theories of organizational grouping, see Mary Ann Spencer Pulaski, *Understanding Piaget: An Introduction to Children's Cognitive Development* (New York: Harper and Row, 1971), 55–56. Claude Lévi-Strauss, in *The Elementary Structures of Kinship* (Boston: Beacon Press, 1969), has argued even more fundamentally that humankind's ability to think of biological relationships in sets of oppositions (man versus woman, mother versus daughter, wife versus sister) provided a system of social exchange that ultimately transformed the state of nature into a culture of civilization. Finally, bridging the primitive and the modern, Emile Durkheim has claimed that the classificatory systems in complex societies continue to be regulated in great part by the intensity of our affective family experiences and our differentiation between types of kinship relationships. For a concise reference, see Chris Jenks, *Culture* (New York: Routledge, 1993), 27.
3. Sergei Eisenstein, "Dickens, Griffith, and the Film Today," in *Film Form: Essays in Film Theory*, ed. and trans. Jay Leyda (New York: Harcourt Brace Jovanovich, 1949), 232.
4. Rudolf Arnheim, *Film as Art* (Berkeley: University of California Press, 1957), 207–8.
5. Eric Barnouw, *Tube of Plenty: The Evolution of American Television*, rev. ed. (New York: Oxford University Press, 1982), 112.
6. Richard Baker, "Film-Style Video," *Videography* 3, no. 10 (October 1978): 81.
7. Stephen A. Booth, "8mm Makes Its Move," *Video* 9, no. 6 (September 1985): 78.
8. Thomas R. Dilts and David Raub, "Fire, Theft, and Loss—How to Protect Your Video Collection," *Video Review* 2, no. 2 (May 1981): 60.
9. Booth, 77.
10. Roy Armes, *On Video* (London: Routledge, 1988), 127–28.
11. Jarice Hanson, *Understanding Video: Applications, Impact, and Theory* (Beverly Hills: Sage, 1987), 16.
12. N. D. Batra, *The Hour of Television: Critical Approaches* (Metuchen, N.J.: Scarecrow Press, 1987), 41.
13. Sean Cubitt, *Videography: Video Media as Art and Culture* (New York: St. Martin's Press, 1993), 32.
14. Armes, 63.
15. Armes, 123.
16. John Belton, "Looking through Video: The Psychology of Video and Film," in *Resolutions: Contemporary Video Practices*, ed. Michael Renov and Erika Suderburg (Minneapolis: University of Minnesota Press, 1996), 65.
17. "It may not be accurate to regard knowledge at the production locus of a commodity as exclusive-

ly technical or empirical and knowledge at the consumption end as exclusively evaluative or ideological. . . . The two poles are susceptible to mutual and dialectical interaction." Arjun Appadurai, "Introduction: Commodities and the Politics of Value," in *The Social Life of Things: Commodities in Cultural Perspective*, ed. Arjun Appadurai (Cambridge: Cambridge University Press, 1986), 41.

18. Douglas Davis, *The Five Myths of Television Power or Why the Medium Is Not the Message* (New York: Simon and Schuster, 1993), 149.

19. Davis, 22.

20. I must admit that this chapter marginalizes alternative representations of race, ethnicity, and class by focusing almost exclusively on TV series about white middle-class families. A word of explanation is in order. To concisely illustrate my primary thesis about the diachronic interpenetrations of home video and domestic television conventions, my methodology has required that I select highly similar family representations from a very specific, necessarily limited range of series acting as a control group with proscribed variables. Although I do touch on sociopolitical and cultural changes that altered the family dynamics of these series somewhat differently, I make no claim that these changes were taking place in other domestic sitcoms or reflecting the real-world experiences of their audiences. I welcome other scholars whose specialty concerns televisual representations of racial and ethnic minorities to fill in the blind spots of my analysis with additional research and commentary.

21. "When radio comedies were adapted to the television medium, critics judged the degree to which the addition of sight to sound enhanced the characters' credibility and the programs' overall sense of intimacy." Lynn Spigel, *Make Room for TV: Television and the Family Ideal in Postwar America* (Chicago: University of Chicago Press, 1992), 157.

22. Certainly African American, Asian American, Latino, and working-class families would probably find it more difficult to identify with the Nelsons, Andersons, and Cleavers than would white middle-class families. At the very least, however, general audiences may have been intrigued with the overall notion of watching the quotidian events of any ordinary family play out week after week—perhaps in the same manner that variegated audiences may identify with a game show champion whose race and ethnicity (and even sexual orientation, in the case of *Survivor*) recede in lieu of the sheer fascination of winning the grand prize.

23. David Morley, "Television: Not So Much a Visual Medium, More a Visible Object," in *Visual Culture*, ed. Chris Jenks (New York: Routledge, 1995), 182.

24. Horace Newcomb, "Toward a Television Aesthetic," in *Television: The Critical View*, ed. Horace Newcomb, 2d. ed. (Austin: University of Texas Press, 1979), 436.

25. Roger Silverstone, *The Message of Television: Myth and Narrative in Contemporary Culture* (London: Heinemann Educational Books, 1981), 67.

26. Michael K. Saenz, "Television Viewing as a Cultural Practice," in *Television: The Critical View*, ed. Horace Newcomb, 5th ed. (New York: Oxford University Press, 1994), 578.

27. John Fiske, *Television Culture* (New York: Routledge, 1987), 22–23.

28. Nina C. Leibman, *Living Room Lectures: The Fifties Family in Film and Television* (Austin: University of Texas Press, 1995), 39.

29. Fiske, 150.

30. Leibman, 56.

31. Leibman, 64.

32. Sol Worth, *Studying Visual Communication*, ed. Larry Gross (Philadelphia: University of Pennsylvania Press, 1981), 181.

33. Spigel, 139.

34. "The ideal sitcom was expected to highlight both the experience of theatricality and the naturalism of domestic life. At the same time that family comedies encouraged audiences to feel as if they were in a theater watching a play, they also asked viewers to believe in the reality of the families presented on the screen." Spigel, 157.

35. M. S. Piccirillo, "On the Authenticity of Televisual Experience: A Critical Exploration of Para-social Closure," *Critical Studies in Mass Communication* 3 (1986): 348.

36. Spigel, 159.

37. Spigel, 165.

38. L. S. Vygotsky, *Mind in Society: The Development of Higher Psychological Processes*, ed. Michael Cole, Vera John-Steiner, Sylvia Scribner, and Ellen Souberman (Cambridge: Harvard University Press, 1978), 94–95.

39. Tony Wilson, *Watching Television: Hermeneutics, Reception, and Popular Culture* (Cambridge: Polity Press, 1993), 38–39.

40. Because the sitcom has been the primary genre in which African American families are represented on television, *The Cosby Show* has received enormous critical attention—both positive and negative—for its similarities to the traditional white middle-class domestic sitcoms discussed

in this chapter. Future studies would benefit greatly from an analysis of the ways in which other popular African American domestic sitcoms such as *The Jeffersons, Good Times,* or *Family Matters* might subvert or parody both home mode conventions and their predominantly white predecessors.

41. *The Adventures of Ozzie and Harriet* (ABC) was first telecast on 3 October 1952 and last telecast on 3 September 1966.
42. Quoted in Gerard Jones, *Honey, I'm Home! Sitcoms: Selling the American Dream* (New York: Grove Weidenfeld, 1992), 95.
43. Quoted in Leibman, 47.
44. *Variety,* 18 September 1964.
45. Quoted in Leibman, 71.
46. Harry Castleman and Walter J. Podrazik, *Watching Television: Four Decades of American Television* (New York: McGraw-Hill, 1982), 76–77.
47. Darrell Y. Hamamoto, *Nervous Laughter: Television Situation Comedy and Liberal Democratic Ideology* (New York: Praeger, 1989), 24–25.
48. Quoted in Arlene Skolnick, *Embattled Paradise: The American Family in an Age of Uncertainty* (New York: HarperCollins, 1991), 126.
49. Quoted in Skolnick, 126.
50. Dan Graham, "Video in Relation to Architecture," in *Illuminating Video: An Essential Guide to Video Art,* ed. Doug Hall and Sally Jo Fifer (New York: Aperture, 1990), 169–70.
51. Graham, 169.
52. Alan Raymond and Susan Raymond, "Filming *An American Family,*" *Filmmakers Newsletter* 6, no. 5 (March 1973): 19–20.
53. Quoted in Anne Roiphe, "Things Are Keen but Could Be Keener," in *An American Family,* ed. Ron Goulart (New York: Warner, 1973), 10.
54. Quoted in Eric Krueger, "An American Family, an American Film," *Film Comment* 9, no. 6 (November–December 1973): 17.
55. See Richard Chalfen's reading of Allan Sekula's description of the "realist" folk myth in *Snapshot Versions of Life* (Bowling Green: Bowling Green State University Popular Press, 1987), 120–21.
56. For a detailed description of documentary in the "observational mode," see Bill Nichols, *Representing Reality: Issues and Concepts in Documentary* (Bloomington: Indiana University Press, 1991), 38–44.
57. Raymond and Raymond, 20.
58. Craig Gilbert, "All in the (Loud) Family," *Television Quarterly* 11, no. 1 (fall 1973): 16.
59. Chalfen, 10–11.
60. Patricia Loud, letter to *The Forum for Contemporary History,* in *An American Family,* ed. Ron Goulart (New York: Warner, 1973), 236.
61. Nichols, 122.
62. Gilbert, 14.
63. Douglas Davis, "Filmgoing/Videogoing: Making Distinctions," in *Video Culture: A Critical Investigation,* ed. John G. Hanhardt (New York: Visual Studies Workshop Press, 1986), 273.
64. Quoted in Raymond and Raymond, 19.
65. Quoted in Melinda Ward, "An Interview with Pat Loud," *Film Comment* 9, no. 6 (November–December 1973): 22.
66. Raymond and Raymond, 21.
67. Krueger, 18.
68. Although the Raymonds shot all of the footage of the Louds, thus bringing to the documentary their own in-camera editorial choices, as the executive producer of the series, Gilbert oversaw the postproduction editing process to shape the footage according to a larger narrative plan he had in mind. Because the Raymonds and Gilbert did not always agree about how each episode should be constructed, I have not taken it for granted that either party has always spoken for the other. Therefore my attributions of Gilbert's intentions derive primarily from his own written statements claiming authority for how the series was structured as broadcast.
69. Davis, *Five,* 25.
70. Graham, 170.
71. Jay Ruby, "Speaking For, Speaking About, Speaking With, or Speaking Alongside: An Anthropological and Documentary Dilemma," *Journal of Film and Video* 44, nos. 1–2 (spring–summer 1992): 50.
72. *The Wonder Years* (ABC) was first telecast on 15 March 1988 and last telecast on 1 September 1993.
73. Miles Beller, review of *The Wonder Years, Hollywood Reporter,* 2 December 1988, 10.
74. Tom Carson, "Sixtiessomething," *Film Comment* 24, no. 4 (July–August 1988): 78.
75. Barry Golson, "A Farewell to Wonder," *TV Guide,* 8 May 1993, 23.
76. In his autobiography, Roland Barthes similarly introduces his past life with a home mode photo

collection, an "image-repertoire" that must, however, be followed by writing to express his truly "productive life." See *Roland Barthes*, trans. Richard Howard (New York: Noonday Press, 1977), 1–42.

77. Executive producer Bob Brush, quoted in Steve Weinstein, "*The Wonder Years* Faces Growing Pains," *Los Angeles Times*, 3 October1989, part 6, p. 4.

78. Fredric Jameson, *Postmodernism, or The Cultural Logic of Late Capitalism* (Durham: Duke University Press, 1991), 70.

79. "The viewer's relationship to this site of cultural construction therefore assumes an increasingly intimate and biographical tone, enabling retrospective periodization of one's life according to collectively shared narrative performances." Michael K. Saenz, "Television Viewing as a Cultural Practice," in *Television: The Critical View*, ed. Horace Newcomb, 5th ed. (New York: Oxford University Press, 1994), 582.

80. Frank Tillman, "The Photographic Image and the Transformation of Thought," *East-West Film Journal* 1, no. 2 (June 1987): 100.

81. *The Real World* (MTV) premiered on 21 May 1992 and continues to run at present.

82. Daniel Cerone, "MTV's Sort-of-Real 'World,'" *Los Angeles Times*, 28 May 1992, F12.

83. Quoted in Diana Moneta, "MTV Shoots *The Real World*," *Videography* 17, no. 5 (May 1992): 40, 42.

84. Steven Mintz and Susan Kellogg, *Domestic Revolutions: A Social History of American Family Life* (New York: Free Press, 1988), 203.

85. Quoted in Cerone, F12.

86. This special aired on ABC on 31 July 1995.

87. This special aired on ABC in May 1995.

88. This special aired on Fox on 18 May 1995.

89. This special aired on ABC on 16 October 1994.

90. This special aired on ABC on 24 May 1995.

91. M. M. Bakhtin, "Forms of Time and Chronotope in the Novel," in *The Dialogic Imagination*, ed. Michael Holquist, trans. Caryl Emerson and Michael Holquist (Austin: University of Texas Press, 1981), 84–85.

92. Robert Stam, *Subversive Pleasures: Bakhtin, Cultural Criticism, and Film* (Baltimore: Johns Hopkins University Press, 1989), 11.

93. Bakhtin, 224–42.

94. Victor Turner, "Acting in Everyday Life and Everyday Life in Acting," in *From Ritual to Theatre: The Human Seriousness of Play* (New York: Performing Arts Journal Publications, 1982), 104–5.

95. For a concise summary of Jauss's "horizon of expectation," see Susan R. Suleiman, "Introduction: Varieties of Audience-Oriented Criticism," in *The Reader in the Text: Essays on Audience Interpretation*, ed. Susan R. Suleiman and Inge Crosman (Princeton: Princeton University Press, 1980), 35–37.

96. John Caldwell, *Televisuality: Style, Crisis, and Authority in American Television* (New Brunswick, N.J.: Rutgers University Press, 1995), 267.

97. *America's Funniest Home Videos* ran through the 1996–1997 season with Bob Saget as host. In the 1997–1998 season, the series was retooled as *America's Funniest Videos* and hosted by Daisy Fuentes and John Fugelsang. My analysis focuses only on the first incarnation, which was far more successful in the ratings.

98. Descriptions of the production process have been derived from intensive research conducted at Vin Di Bona Productions in Los Angeles from August 1995 through March 1996, including interviews with producers, notes from staff meetings, and observation at editing suites and studio tapings. I must stress the importance of doing actual ethnographic investigations to understand the production process of any series in accurate detail. In the case of *America's Funniest Home Videos*, published analyses of its operations too often depend entirely on formal readings of its broadcasts, whose elisions of behind-the-scenes operations and taken-as-evident assumptions frequently cause analysts to jump to erroneous conclusions.

99. Richard Goedkoop, "The Game Show," in *TV Genres: A Handbook and Reference Guide*, ed. Brian G. Rose (Westport, Conn.: Greenwood Press, 1985), 288.

100. Goedkoop, 287.

101. See, for instance, Laurie Ouellette, "Camcorder Dos and Don'ts: Popular Discourses on Amateur Video and Participatory Television," *Velvet Light Trap* 36 (fall 1995): 34–44; and Steve Fore, "America, America, This Is You! The Curious Case of *America's Funniest Home Videos*," *Journal of Popular Film and Television* 21, no. 1 (spring 1993): 37–45.

102. See Conrad Lodziak on the "power-of-television thesis" in *The Power of Television: A Critical Appraisal* (London: Frances Pinter, 1986), 119–20.

103. Lodziak, 121.

104. Barbara Bernstein, interview by author, 6 September 1995.

105. Quoted in Daniel Cerone, "The Candid Camcorder: *America's Funniest Home Videos* Taps Our Living Rooms for Some Home-Grown Humor," *Los Angeles Times,* 14 February 1990, F1.

106. David Hajdu, "The Camcorder," *Video Review* 10, no. 1 (February 1990): 35.

107. Cecelia Tichi, *Electronic Hearth: Creating an American Television Culture* (New York: Oxford University Press, 1991), 137.

108. Bernstein interview.

109. Stuart Hall, "Notes on Deconstructing 'the Popular,'" in *People's History and Socialist Theory,* ed. Raphael Samuel (London: Routledge, 1981), 227–39.

110. Bernstein interview.

111. Pierre Bourdieu, *Photography: A Middle-Brow Art,* trans. Shaun Whiteside (Stanford: Stanford University Press, 1990), 27.

112. Mary Conley, interview by author, 19 December 1995.

113. Jeffrey Cole, *The UCLA Television Violence Monitoring Report* (Los Angeles: UCLA Center for Communication Policy, 1995), 47.

114. Cole, 16.

115. Richard Dienst, *Still Life in Real Time: Theory after Television* (Durham: Duke University Press, 1994), 166.

5. The Video-in-the-Text

1. For example, an article entitled "The Great Face-Off: Video vs. Film: Two Experts Exchange Fire" features D. A. Pennebaker and John Reilly debating the merits of film and video for a proper documentary practice. A defender of film, Pennebaker believes that the medium's limitations (short reels, laboratory expenses, need for strong lighting) are better suited for critical documentaries, which require advanced planning, deliberate in-camera editing, and thoughtful decisions about content during the act of shooting. For Reilly, a defender of video, the main object is to shoot as much as possible, preferring not to shape the material with a priori decisions constrained by reel changes, and thus capturing the unplanned moments of life. *Video Review* 5, no. 11 (February 1985): 28–30.

2. "Driven by the nightmare of hearing the phrase 'we could just as well have done it on tape,' the film community made its internal power accommodations and promoted its mystique of quality in order to survive in the higher-budget ends of the industry." Charles Eidsvik, "Machines of the Invisible: Changes in Film Technology in the Age of Video," *Film Quarterly* 42, no. 2 (winter 1988–1989): 22.

3. See M. M. Bakhtin, *The Dialogic Imagination* (Austin: University of Texas Press, 1981), 270–72 (unitary language), 342–44 (authoritative discourse).

4. Robert Stam, *Subversive Pleasures: Bakhtin, Cultural Criticism, and Film* (Baltimore: Johns Hopkins University Press, 1989), 52–53.

5. Lew McCreary, "Film and Tape: Mixing Oil and Water," *Filmmakers Newsletter* 9, no. 6 (April 1976): 29.

6. James B. Brandt, "Video Assist: Past, Present, and Future," *American Cinematographer* 72, no. 6 (June 1991): 93.

7. Chris Darke, "Sibling Rivalry: Cinema and Video at War," *Sight and Sound* 3, no. 7 (July 1993): 28.

8. This chapter is concerned first and foremost with cinematic narratives portraying amateur and home video, whose numbers increase monthly. If we include representations of surveillance video, industrial video, and television, this number increases exponentially. Discussions of nonnarrative experimental films that appropriate video would require another study altogether.

9. Gérard Genette, *Narrative Discourse: An Essay in Method,* trans. Jane E. Lewin (Ithaca, N.Y.: Cornell University Press, 1980), 232–33.

10. Bakhtin, 304.

11. "The technology underlying stylistic technique is interpreted as such only when it differs from the interpretant's time-based, ideologically limited conception of the technological norm. Such is the case with the adoption of 'new' technology by the film industry. The spectator will consider this technology significant because it differs from the norm to which he or she is accustomed." Jeremy Butler, "A Compendium of Stylistic Interpretation," *Film Reader* 5 (1982): 288.

12. "A figure functions only when it is observed to function, only when it stands in the way of an automatic movement across signs. If, as usual, nothing halts us at the level of perception, the next potential figural work occurs at the level of narration." Dudley Andrew, "On Figuration," *Iris* 1, no. 1 (1983): 126.

13. Sean Cubitt, *Videography: Video Media as Art and Culture* (New York: St. Martin's Press, 1993), 164–65.

14. Christian Metz, "The Imaginary Signifier (Excerpts)," in *Narrative, Apparatus, Ideology: A Film Theory Reader,* ed. Philip Rosen (New York: Columbia University Press, 1986), 244–78.

15. Stephen Heath, "The Cinematic Apparatus: Technology as Historical and Cultural Form," in *The*

Cinematic Apparatus, ed. Teresa de Lauretis and Stephen Heath (New York: St. Martin's Press, 1980), 2–3.

16. André Bazin, "The Ontology of the Photographic Image," in *What Is Cinema?* vol. 1, trans. and ed. Hugh Gray (Berkeley: University of California Press, 1967), 9–16.

17. "Images can cue such hypotheses without the real camera's ever having been in any such position, as animators have known for decades. All that people need in order to construct the schema called 'camera,' it seems, are some assumptions about how photographic images are produced." David Bordwell, *Narration in the Fiction Film* (Madison: University of Wisconsin Press, 1985), 119.

18. My discussion of existential phenomenology and its value for media studies is indebted to Vivian Sobchack, *The Address of the Eye: A Phenomenology of Film Experience* (Princeton: Princeton University Press, 1992).

19. Stanley Fish, *Is There a Text in This Class? The Authority of Interpretive Communities* (Cambridge: Harvard University Press, 1980), 14–17.

20. For a concise summary of Jauss's "horizon of expectation," see Susan R. Suleiman, "Introduction: Varieties of Audience-Oriented Criticism," in *The Reader in the Text: Essays on Audience Interpretation*, ed. Susan R. Suleiman and Inge Crosman (Princeton: Princeton University Press, 1980), 35–37.

21. Janet Staiger, *Interpreting Films: Studies in the Historical Reception of American Cinema* (Princeton: Princeton University Press, 1992), 46.

22. Darke, 27.

23. Patricia R. Zimmermann, *Reel Families: A Social History of Amateur Film* (Bloomington: Indiana University Press, 1995), 157.

24. Roger Odin, "A Semio-pragmatic Approach to the Documentary Film," in *The Film Spectator: From Sign to Mind*, ed. Warren Buckland (Amsterdam: Amsterdam University Press, 1995), 227–35.

25. Odin, 222.

26. This summary is indebted to Vivian Sobchack's reading of Meunier in "Toward a Phenomenology of Nonfictional Film Experience," in *Collecting Visible Evidence*, ed. Jane M. Gaines and Michael Renov (Minneapolis: University of Minnesota Press, 1999), 241–54.

27. Edward Branigan, *Point of View in the Cinema: A Theory of Narration and Subjectivity in Classical Film* (New York: Mouton, 1984), 209–10.

28. David Bordwell, *Making Meaning: Inference and Rhetoric in the Interpretation of Cinema* (Cambridge: Harvard University Press, 1989), 162–63.

29. "The image can become specifiable in its relationship as a production through awareness of the camera and its properties as a mechanism, thus counteracting the omniscience of the point of view by giving limits to the method of access to the scene. In this way, the viewpoint becomes relatable to a causal agency, readable in a materialist sense rather than as a generalized ideal access (in effect unread through the transparency of its inscription)." Malcolm Le Grice, "Problematizing the Spectator's Placement in Film," in *Cinema and Language*, ed. Stephen Heath and Patricia Mellencamp (New York: University Publications of America, 1983), 57.

30. Rosalind Krauss, "The Impulse to See," in *Vision and Visuality*, ed. Hal Foster (Seattle: Bay Press, 1988), 55, 58.

31. Nick Browne, "The Spectator-in-the-Text: The Rhetoric of *Stagecoach*," in *Film Theory and Criticism: Introductory Readings*, ed. Gerald Mast, Marshall Cohen, and Leo Braudy, 4th ed. (New York: Oxford University Press, 1992), 210–25.

32. See Metz, 253–55.

33. See Genette, 29–32.

34. Maureen Turim, *Flashbacks in Film: Memory and History* (New York: Routledge, 1989), 1–2.

35. See, for example, Heike Klippel, "Film and Forms of Remembering," *Iris* 17 (fall 1994): 119–35.

36. Jean Piaget and Bärbel Inhelder, *The Psychology of the Child* (New York: Basic Books, 1969), 80.

37. For a concise description and excellent application of Freud's "screen memory," see Marita Sturken, "The Politics of Video Memory: Electronic Erasures and Inscriptions," *Resolutions: Contemporary Video Practices*, ed. Michael Renov and Erika Suderburg (Minneapolis: University of Minnesota Press, 1996), 1–12.

38. Genette, 86–112.

39. See Genette, 95. Genette qualifies the possibility, however, that written language can actually express slow motion within narrative action.

40. Marsha Kinder, "The Subversive Potential of the Pseudo-Iterative," *Film Quarterly* 43, no. 2 (winter 1989–1990): 3–16.

41. See sections 28, 40, and 81 of Roland Barthes, *S/Z: An Essay*, trans. Richard Miller (New York: Noonday Press, 1974).

42. For more on "orientational metaphors," see George Lakoff and Mark Johnson, *Metaphors We Live By* (Chicago: University of Chicago Press, 1980), 14–21.

43. Stanley Milgram, "The Image-Freezing Machine," *Psychology Today* (January 1977): 108.

Index

Baudry, Jean-Louis, 21, 24, 171
Bazin, André, xi, xiii; on cinema, 21; death mask and, 24; on photographic realism, 172; on plastic arts, 94–95, 115; on postmortem images, 95
Before They Were Stars, 141
Beller, Miles, 133
Belton, John, 10, 102
Benjamin, Walter, 78
Benning, Sadie, 20, 56
Berko, Lili: on video, 13–14
Berliner, Allan, 187, 188
Bernstein, Barbara, 156, 159, 160
Bertolucci, Bernardo, 202
Big Chill, The, 167, 198
Black, Carol: *Wonder Years* and, 132
Blade Runner, 189
Boddy, William: on television, 21–22
Bordwell, David, 181
Bourdieu, Pierre, 64, 84; on classificatory schemes, 25; on cultural practice, xiv, 28; distinction and, 28; on habitus, xvii, 35, 54, 70; on photographic practice, 52, 70; on social class, 55, 81; sociology of, 28, 29; theory of practice of, 54, 56, 67
Brady Bunch, The, 112, 125
"Brady Bunch Home Movies," 148
Brakhage, Stan, xviii, 67, 69, 88; avant-garde and, 73, 74, 75; on home moviemaking, 210n15
Branigan, Edward, 181
Brecht, Bertolt, 78
Browne, Nick: on *Stagecoach,* 182
Bundle of determinations, 177, 178
Bunim, Mary-Ellis, 141, 143, 144
Burger, Peter: on avant-garde, 76
Burns, George, 108, 114
Butler, Brett, 146–47

Cain, Dean, 145, 146
Caldwell, John: on television/video tape, 152
Calendar, 30, 175
Camcorders, 33, 42, 53, 90, 102, 147, 155, 158, 159, 168, 171, 174, 176, 178, 184, 191, 195, 196, 198, 199
Camera Lucida: Reflections on Photography (Barthes): quote from, 64
Cameras, 75, 217nn17, 29; film, 117, 169, 171, 172, 174, 175, 183; Super-8, 41; video, 88, 169, 171, 185
Candid Camera, 131, 147, 152, 154
Capitalism, 9; communications technologies and, 18; functionalism/strain theory and, 57; Kodak culture and, 49
Capitalism and Communication (Garnham), 89
Carroll, Noël, 4
Carson, Tom, 133
Causality, 54; expressive, 52; historical, 38; mechanistic, 2, 4, 34, 205n4; problems of, 2–6; structural, 18; symptomatic, 3, 15, 18
Chalfen, Richard, 52; ahistoricism of, 37, 39; on amateurs, xvi, 37; home mode and, 34–35, 36–39, 41, 43, 44, 49, 59, 106–7; on home movies, 40, 42, 94; on Kodak culture, 50; on photographers/photography, 123; scholarship

of, 35; on snapshots, 94; on social functions, 39; transhistorical formalism of, 39
Children's Television Act (1990), 161
Chronotope, xxi, 113, 149–52
Cinema, 169; allegories of, 87; discursive rules of, 168; event videographers and, 80; hegemony of, 164, 166; institutionalization of, 166; narrative, xxi, 16, 181; television and, 12, 99, 101, 164, 166, 167, 168; video and, 12, 14, 99, 101, 164, 166, 168
Cinematic narrative, xiv, xxi, 181, 189, 197, 200
Cinema verité, 121, 124, 128, 130, 142, 155
Citron, Michelle, 53, 56
Clarke, Wendy, 11
Classical film theory, xi, xii, xiii, 24
Cocker, Joe, 134, 136
Cognitive mapping, 189–91
Cohen, Maxi, 11
Cole, Jeffrey, 161, 162
Comolli, Jean-Louis, xxiii
Congressional Record: Sarnoff in, 22
Conley, Mary, 161
Consumption, 82, 83, 88; home mode and, 57; production and, 58, 102, 103
Cops, 141
Corrao, Lauren, 141
Cosby Show, The, 112, 213n40
Crawford Memorials, 93, 95
Credit sequences, 111, 133, 134, 136, 175; analyzing, 108–9
Crimson Tide: home video in, 178
Critical Theory of the Family (Poster), 46
Cubitt, Sean: on video, 9, 23–24, 100
Cultural capital, 29, 55, 82, 84

Darke, Chris, 163, 177
Daughter Rite (Citron), 53
Davis, Douglas: on *American Family,* 124; on electromagnetic tape, 104; on television, 103, 104, 131; on video, 103, 104
Death Watch, 13, 181
De Certeau, Michel, 20, 81
Defending Your Life, 200
Deleuze, Gilles, xxii, 24
Deren, Maya, xviii, 67; avant-garde and, 73, 74, 75
Derrida, Jacques, xiii, 1, 28
Determinism, 19; determination and, 20; mechanistic, 3, 34, 53; technological, xii, xv, 2, 3, 4, 20, 39, 78, 102, 104
Dialogic Imagination, The (Bakhtin), 149
Dialogism, 29–30, 156, 177
Di Bona, Vin, 152–57, 215n98
Dienst, Richard, 162
Dilks, John: talking tombstones and, 93
Discourses, 34, 49, 50, 53, 92, 181, 190; amateur-film, 52; commercial, 36, 52; epiphenomenal, xvi, 54; epistemological value of, 25; progressive media, 79, 96, 97–98
Distinction: A Social Critique of the Judgements of Taste (Bourdieu): quote from, 64
Documentaries, 69, 142, 145, 170, 202; cultural function of, 68; home mode and, 121–32; as

mode of practice, 68; observational mode of, 123; professional identity and, 178; sitcoms and, 121, 144; subjectivity of, 124; taxonomy of, 68; television and, 131

Dog Star Man, 71

Domesticity, xvii, 57, 141, 145, 188; historical transformations of, 135–36; in home movies, 45; professional identities and, 149; theatricality of, 143, 213n34

Dominant ideology, xvii, 28, 29, 35, 39, 48, 49, 50, 51, 56, 66, 77, 164

Donna Reed Show, The: white middle class and, 105

Down and Out in Beverly Hills, 177, 195, 203

Dramedy: home mode in, 132–41

Dubois, Philippe, xi, xiii

Durkheim, Emile: primitive/modern and, 212n2

Eastman Kodak Company. *See* Kodak

Economic capital, 29, 55, 73, 82, 84

Economic relations, 21, 58; functional modalities and, 67–72

Egoyan, Atom, 165–69, 174, 191, 192, 193, 199, 201; cinema of, 29–30, 167, 195; Wenders and, 163

Eisenstein, Sergei, xii, 24, 98

Either/or binarisms, xiii, xx, 8, 10, 13, 15, 34, 39, 69

Electronic Hearth (Tichi), 158

Entertainment Tonight, 155

Enzensberger, Hans, 7, 78, 210n29

Epistemology, 24–25, 62; realist/formalist binaries of, 10

Esland, Geoff: on home video/event video, 81

Ethnographies, 80, 90

Event videographers, 16, 64; clients and, 83, 89, 91; equipment for, 65, 84, 87; family photography and, 82; funeral directors and, 92; home movies and, 80; home video and, 82, 88; marketing by, 81; professional/amateur, 79–89; professionalism of, 65, 80, 83; symbolic capital and, 84

Event videography, xviii, 84, 95, 96; commercial nature of, 90; professionals, xix, 90; as semi-bourgeois career, 83

Event videos, 89–96, 103; expertise and, 90–91; home videos and, 80, 81, 83, 88; industrial video and, 83; professional value of, 83; simulation of, 88; television and, 65–66, 80, 83

Exotica, 29–30, 184

Falling Down: traffic jam opening, 189

Familialism, 29, 47, 110–11, 120; critique of, 138; ethnicity/geography/technology and, 188; home video and, 52; ideology of, 35, 38, 39, 44, 45, 46, 130–31, 177; individualism and, 126; power dynamics of, 98

Families, 143; changes in, 48; conventional ideas of, 125; cultural images of, 122; idealized, 131; metaphors of, 98; sitcoms and, 118–19; television, 101, 106, 120, 123. *See also* Nuclear families

Family Album, The, 187

Family albums, 128, 145

Family folklore, xviii, 36, 61, 150

Family life, 135; home mode and, 111; home movies and, 45; images of, 106, 109, 137; sitcoms and, 111

Family Matters, 214n40

Family photography, 82, 101, 108, 111, 159; sitcoms and, 105; television and, 152

Family resemblances, xx, 97, 101, 109–13, 166, 212n2

Family romance, 191–94

Family Ties, 112

Family Viewing, xxi, 191, 193, 195, 201, 203; Oedipal rivalries in, 30, 163

Father Knows Best, 105, 122, 135

Feuer, Jane: on morning news programs, 22

Field, 18, 54–58; habitus and, 54

Film, 103; amateur, 33, 45, 52, 208n24; home video and, xxi, 177; identity and, 163; as masculine, 98–99; ontology of, 165; television and, xvi, 19, 104, 167; video and, 12, 13, 39–44, 100, 102, 163, 164–67, 169, 176, 187

Film as Art (Arnheim), xii

Filmic narratives, 163, 164, 167, 175, 178, 184, 188

Film-souvenir, 72–73, 209n11

Film theory, xi, xii, xiii, 24

Film-to-tape transfers, 135, 166–67

Fish, Stanley, 176

Fisher Price: pixelvision and, 20

Fiske, John: on television/real life, 107

Flashbacks, 118, 184

Flow: concept of, 15

Frankfurt school, 78

Frank's Place, 132

Freeze-frames, 43, 145

Freud, Sigmund, 184, 191, 192

Fuentes, Daisy, 215n97

Fugelsang, John, 215n97

Full House, 119, 154

Functionalism, 18, 19, 28, 56; teleology and, 113

Functional modality: economic relations and, 67–72; home mode and, 72–73

Funeral directors: event videographers and, 92

Funt, Alan, 154

Fun with Ken and Kato Chan, 152

Garnham, Nicholas, 78, 89

Generations (Bolling), 9

Genette, Gerard, 183, 185, 186; metadiegesis of, 168; on narrative action, 217n39; on pseudo-iterative, 187

Genres, 68, 101, 211n37

Gilbert, Craig, 132, 136, 139, 141–42; on American families, 122; *American Family* and, 121; cinema verité and, 130; documentary by, 123, 144; home mode and, 129; on Louds, 122, 124, 125–26, 129–30

Global Village, 7

Godard, Jean-Luc, 163, 189

Godfrey, Arthur, 154

Golson, Barry, 133

Goodman, Nelson: on theoretical objects, 23

Graham, Dan, 122, 123, 131

Gramsci, Antonio: hegemony and, 28, 30

Grand Syntagmatique, xii

Grierson, John, 68, 71
Guattari, Félix, xxii
Guerrilla Television (Shamberg), 7

Habenstein, Robert: on funeral service industry, 92
Habitus, 48–50, 70; described, 35, 54–58; field and, 54; home mode, 56, 189; medium specificity and, 55
Hall, Stuart, xxiii, 159
Halleck, DeeDee: on home movies, 88
Heath, Stephen, 21, 172
Hegemony, 25–26, 28, 30, 31, 49; class, 67; medium specificity and, 25
Henry: Portrait of a Serial Killer, 185, 202, 203
Hermeneutic methodologies, 87, 190
Hershman, Lynn: on new autobiography, 11
Heteroglossia, 156, 166, 169
Heuristics, 22–24
Hirsch, Eric, 58
Holliday, George, 131, 152, 159
Home for the Holidays, 191, 200
Home mode, 34–35, 53, 55, 66, 113; bourgeois realism of, 29; conservatism of, 48, 76; critiquing, 44, 67, 138; cultural functions of, 88, 106–7; cultural reproduction and, 48–50; defining, 36–39, 58; discourse of, 49, 50; families, 44–48, 111, 129; folk myth and, 123; functional taxonomy of, 59–63; ideologies of, 36, 77; intersubjective relations of, 89, 90–91; persistence of, xvi, 48; reality in, 141–49; ritual functions of, 36; self-determination and, 89; social function of, 78; static model of, 38; symbolic strategies of, 70; technologies of, 39–44; television and, 100, 105–7, 109, 111, 125, 144, 148–49, 152–62; video and, 50, 149
Home mode artifacts, 50, 127, 128, 148; autotelic, 68; content/form of, 49, 60; function of, 55–56, 60–61; sitcoms and, 110; social order and, 52; as sociological data, 37
Home moviemakers, 40, 41, 210n15
Home movies, xvi, xvii, xviii, 35, 39, 48, 58, 93, 101, 105, 106, 109, 123, 127, 128, 131, 149, 150, 159; duplication of, 42, 53, 135; family relations and, 45; home video and, 40, 43, 134, 154; institution of, 78; as Kodak culture, 49; limitations of, 41; narrative content of, 41; nostalgia of, 139; production of, 41; reception of, 42; representations in, 73, 138; residual pleasures of, 120; television and, 117, 132; video and, 42
Home video, xiv, 29, 48, 58, 66, 68, 93, 118, 141, 155, 156, 177, 178, 194; analysis of, xv, 34, 35, 73, 84; autobiographical functions of, 62; context/relation in, xxiii; cultural prohibitions and, 44; determinations of, 39; discourses on, 34, 52, 53, 190; divergent dialect of, xxii; equipment for, 30, 84; as ethnographic practice, xv; event video and, 80, 81, 83, 88; family representations of, 101; as heteroglot texts, 157; home movies and, 40, 43, 134, 154; parody of, 172–73; predominance of, 51; production/reception of, 42–43; recording on, 27, 42; specificities/cultural codes of, xxii; taxonomy of, xviii, xxii; television

and, xx, xxi, 65, 97, 101, 104–9, 111, 135, 147, 148, 150, 152, 157–58, 161; transparency of, 42
Home video artifacts, 43, 48, 159
Home videographers, 81, 85; avant-garde and, 70; event videographers and, 82; family photography and, 82; technology for, 65
Horizon of expectations, 151, 176, 177
Hybridity, 16–22

Identification, 73, 107; cinematic, 171–72, 183; modes of, 180; spectatorial, 112; television and, 147
Identities, 61; commercial, 58; cultural, 72; film/photography and, 163; human, 62; individual/group, 60; personal/public, 60; political, 72; professional, 146, 149, 178; public/commercial/private, 60; self-, 9, 16, 80; spiritual, 37
Ideological state apparatuses (ISAs), 18
Ideology, 21, 26–28, 52, 57, 67; aesthetic, 104; familial, 27, 38, 39, 47; social function and, 27; technology and, 26; transformation of, 43
Ideology thesis, 48–54; advertising/manuals/popular entertainment and, 50; mechanistic determinism and, 53; as progressive media doctrine, 51; social order and, 52
Imaginary, 16–22
Indeterminacy, aesthetic of, 12–16
Individualism, 39, 126, 131
Industrial mode, 36, 66, 67, 83, 178
Inherent properties, 5, 9–12, 104
Installations, video, 6–7
Internet, 8, 13, 24
Interpretive communities, 176, 177
I've Heard the Mermaids Singing, 175, 198
I Witness Video, 152

Jade: videotapes in, 185
James, David, 74; on allegories of cinema, 87; on industrial modes of production, 88
Jameson, Fredric, xiii, 139, 206n31; video metaphors and, 25; on video/postmodernism, 15
Jauss, Hans Robert: on chronotopes, 151; on cognition, 176; horizon of expectation and, 151, 215n95, 217n20
Jenks, Chris, 22–23
Jimmy Hollywood, 168, 200
Johnson, Randal, xxiii, 22

Kellogg, Susan, 46–47
Kennamer, Paul D.: on videographers, 209n2
Khanjian, Arsinée, 175
Kinder, Marsha, 187
King, Barry: on photoconsumerism, 208n23
King, Rodney: video of, 152, 159
Kinship, 37, 47, 48, 57, 143, 191; analogies of, 98; portrayal of, 106; taxonomies of, 98
Kodak: commercials by, 51, 52, 116–17, 118; marketing strategies by, 38; monopoly by, 50; Nelsons and, 115; *Ozzie and Harriet* and, 107, 108; video and, 99
Kodak culture, 49, 50
Kopytoff, Igor, 209n45
Kracauer, Siegfried: films and, xi

106, 109, 118, 127, 128, 131, 144, 145, 157. *See also* Photography
Snapshot Versions of Life (Chalfen), 36, 207n10
Sobchack, Vivian, 69, 209n11, 217n26; on film material existence, 173–74
Social capital, 73
Society of Professional Videographers, 84, 209n2
Soft determination, xvi, 16–22
Sony Portapak, 1, 6, 7–8
Speaking Parts, 179–80, 199, 201
Specificity, xii, 1–2, 15, 19, 72, 172; cinematic, 177; exploration of, xviii; material, 9; medium, 22, 37; modernist pursuit of, 12; revised model of, 164
Spectatorship, 182; cinema, 21, 30; hybrid, 173–80; phenomenology of, 167
Spigel, Lynn, 109, 110, 111
Stagecoach: Browne on, 182
Stealing Beauty, 201–2
Strain theory, 56, 57
Structuralism, xxiii, 18, 54
Structured-in-dominance, 18, 69, 71, 72, 156
Sturken, Marita, 1–2, 26
Subjectivism, 54, 91, 181
Super-8 movies, 17, 33, 127, 132, 135, 148
Surveillance video, 141, 178, 179, 182, 194
Symbolic capital, xix, 29, 84

Talking tombstones, 93
Tamblyn, Christine: on marketing campaigns, 53
Taxonomy: binary, 104; constructing, 67; functional, 69, 71; modal, 68, 69; video, 165
Techno-aesthetics, xii, 2–6, 16, 30, 31, 67, 74, 87, 103, 104
Technology, xii, 12, 19, 67; aesthetics and, 1, 2; amateur, 38, 39, 56; communication, 7, 18, 20, 30; digital, xxiv, 17; domesticated, 51; electronic, 100; film, xiii, xxii, 174; home mode, 39–44; home video, 17, 134; ideology and, 26; media, xxiv, 4, 8, 19, 20, 22; motion picture, 117, 207n10; old/new, 17, 78; photographic, 38; professionalism and, 85; rethinking, 16–22, 44; television, 102. *See also* Video technology
Telefilms, 22, 26, 102
Television, 40, 194; analysis of, xx; cultural functions of, 4, 106–7; deterministic theories of, 6; discursive strategies of, 109; domestic, 104–9; evolution of, 98; as feminine, 98–99; as flow, 24; hegemony of, 79, 124, 159; home mode and, 100, 105–7, 109, 111, 125, 144, 148–49, 152–62; home movies and, 117, 132; home video and, xx, xxi, 65, 97, 101, 104–9, 111, 135, 147, 150, 152, 157–58, 161; as homogenizer, 155; identification and, 147; ideological power of, 156; live, 6, 21–22, 26, 102; participatory, 155; reality-based, 67, 103, 104, 105, 107, 141–49, 152, 153; self-reflexive, 10; social reality and, 109; as transmission, 102; video and, 6, 8, 14, 15, 17, 23, 29, 51, 79, 97, 99–105, 141, 145, 147, 156, 162, 164, 165
Theory of practice, 54, 56
Theory of Practice (Bourdieu), xiv
Thicke, Alan, 155

Thigh Line Lyre Triangular, 74
Things to Do in Denver when You're Dead, 191
This Is Your Life, 147
Tichi, Cecelia: on television/modern life, 158
Time-space relations, 149, 150–51, 180
To Die For, 170, 197
*Totally F***ed Up*, xxi, 168, 195, 196, 197; home videotapes in, 190–91
True Love, 181
Turim, Maureen: reflection theory and, 14
Turner, Victor, 24, 75, 150
TV Bra (Paik), 6
TV Chair (Paik), 6
TV Garden (Paik), 6
TV Guide: on Louds, 121; on *Ozzie and Harriet*, 114
"TV Laughs at Life," 146–47
"TV Nation," 51, 77
"TV's All-Time Funniest Christmas Moments," 146
"TV's All-Time Funniest Sitcom Weddings," 146
TV's Bloopers and Practical Jokes, 152
Tyler, Liv, 201, 202

UCLA Center for Communication Policy, 161
UCLA Television Violence Monitoring Report, The: on children/violence, 161
Unsolved Mysteries, 153
Until the End of the World, 163, 191, 203

VCRs, 40, 42–43, 102, 103, 118, 158, 160, 177, 201; controls on, 120; sales of, 99
Veblen, Thorstein: on consumption, 82
Vertov, Dziga: avant-garde media and, 77
Video: atheoretical/asocial, 14, 17; as bastard medium, 100; cinema and, 12, 99, 101, 164, 166, 167, 168; community, 7; critics of, 165; defenders of, 165; digital code of, 40; diversity of, 5, 166; film and, 12, 13, 39–44, 100, 102, 163, 164–67, 169, 187; forms of, 6, 99–100, 173, 174; history of, xv, 1, 2, 12, 31–32, 99, 100; home mode and, 50, 149; iconographic value of, 167–68; as intermediary, 187; liminal status of, 14; as modernist art, 26; parent forms of, 99; properties specific to/inherent in, 11; protean status of, 13–14; representations of, 167; as solipsistic medium, 196; television and, 6, 8, 14, 15, 17, 23, 29, 51, 79, 97, 99–105, 141, 145, 147, 156, 162, 164, 165; video as, xv, 9–12
Videocassettes, 13, 36, 40, 95, 167
Video equipment, 36, 99, 152; prosumer, xix, 84, 85
Videofreex, 7
Videographers, 90, 170, 211n35; agency/production and, 91; male/professional, 92; professional association of, 64; skills of, 91
Videography: amateur, 36; expansion of, 64; funeral, 92, 95
Videography magazine, 84, 85
"Video in Search of a Discourse" (Berko), 13
Video-in-the-text (VIT), xxi, xxii, 167–73, 189, 190, 201, 202; amateurs and, 170; autobiographical impulse of, 197; discourse of, 169, 177, 181; duality/ambiguity of, 174; function of, 183, 184, 186; heteroglossia of, 169; historical/imaginary

existence of, 173; home mode, 191; as hypertext, 169; imaginary apparatus of, 171, 174; metaphysical/psychological nature of, 172; narratology of, 180–88; objective features of, 173; readings of, 176–77; spectator/subject and, 164, 182

Video Maker, 53, 90

"Videor" (Derrida): quote from, 1

Video Review, "Shootoff" competition by, 53

Videotapes, 10, 60, 62, 79, 85, 93, 102, 103, 104, 135, 198, 200; cost of, 41; sound and, 5; Super-8 movies and, 132; television and, 165

Video technology, xxii, 2, 5, 6, 35, 40, 85, 105, 163, 164, 171, 172, 173, 181; filmmaking and, 13; mobilizing power of, xix, 79; proliferation of, 9–10, 33

VIT. *See* Video-in-the-text

Voyeurism, 131, 147, 152, 182, 193, 203; exhibitionism and, 201

Vygotsky, L. S., 111

Wedding photographers, 29

Wedding Video Expo, 210n33, 211n64

Wedding videographers, 84, 90, 199, 210n33

Wedding videos, xix–xx, 89–90, 211n44; amateur/professional, 65; production of, 90, 91; women and, 91–92, 211n37

Weekend: traffic jam in, 189

Welcome to the Dollhouse, 200–201

Wenders, Wim, 163

Weston, Kath: on domestic patterns, 47

"What is Cinema?" (Bazin), xi, xiii

Williams, Raymond, 3; on communications technology, 4, 20; on continuity, 17; on determination, 20; on emergent/residual practices, 18; flow and, 15; on home mode artifacts/social order, 52; mobile privatization and, 205n5; on production process, 31; residual/emergent tendencies and, 17; on soft determination, xvi; on technology, 4; on television, 4, 21

Window Water Baby Moving, 74

Wings of Desire, 163

Wiseman, Fred, 69

Wittgenstein, Ludwig: on family resemblances, 97

Wonder Years, The, xxi, 101, 113, 145; action/reaction and, 142; credit sequence for, 133, 134; domestic themes of, 133; dramedic structure of, 134, 136; home mode and, 132–41; home movie aesthetic of, 142–43; nostalgia of, 136, 138, 139, 150; personal recollections in, 151; popular consciousness and, 140–41

"World's Funniest Wedding Disasters, The," 147

Worth, Sol, 63, 70, 109

Zimmermann, Patricia, 51, 177; on amateur-film discourse, 52; on equipment, 85; home mode and, 34–39, 44; on nuclear families, 44–45; on professional/amateur filmmaking, 85; reel families of, 46

JAMES M. MORAN is a teacher and writer based in Los Angeles. He received a Ph.D. from the University of Southern California's School of Cinema–Television and is currently on the faculty of the visual and media arts department at Emerson College in Los Angeles.